Lecture Notes in Mathematics 2036

Editors:
J.-M. Morel, Cachan
B. Teissier, Paris

For further volumes:
http://www.springer.com/series/304

Volker Mayer • Bartlomiej Skorulski
Mariusz Urbanski

Distance Expanding Random Mappings, Thermodynamical Formalism, Gibbs Measures and Fractal Geometry

 Springer

Volker Mayer
Université Lille 1
Département de Mathématiques
59655 Villeneuve d'Ascq
France
volker.mayer@math.univ-lille1.fr

Mariusz Urbanski
University of North Texas
Department of Mathematics
Denton, TX 76203-1430
USA
urbanski@unt.edu

Bartlomiej Skorulski
Universidad Catolica del Norte
Departamento de Matematicas
Avenida Angamos 0610
Antofagasta
Chile
bskorulski@ucn.cl

ISBN 978-3-642-23649-5   e-ISBN 978-3-642-23650-1
DOI 10.1007/978-3-642-23650-1
Springer Heidelberg Dordrecht London New York

Lecture Notes in Mathematics ISSN print edition: 0075-8434
                              ISSN electronic edition: 1617-9692

Library of Congress Control Number: 2011940286

Mathematics Subject Classification (2010): 37-XX

Printed on acid-free paper

Springer is part of Springer Science+Business Media (www.springer.com)

Preface

In this book we introduce measurable expanding random systems, develop the thermodynamical formalism and establish, in particular, exponential decay of correlations and analyticity of the expected pressure although the spectral gap property does not hold. This theory is then used to investigate fractal properties of conformal random systems. We prove a Bowen's formula and develop the multifractal formalism of the Gibbs states. Depending on the behavior of the Birkhoff sums of the pressure function we get a natural classifications of the systems into two classes: *quasi-deterministic* systems which share many properties of deterministic ones and *essential* random systems which are rather generic and never bi-Lipschitz equivalent to deterministic systems. We show in the essential case that the Hausdorff measure vanishes which refutes a conjecture of Bogenschütz and Ochs. We finally give applications of our results to various specific conformal random systems and positively answer a question of Brück and Büger concerning the Hausdorff dimension of randomJulia sets.

Acknowledgements

The second author was supported by FONDECYT Grant no. 11060538, Chile and Research Network on Low Dimensional Dynamics, PBCT ACT 17, CONICYT, Chile. The research of the third author is supported in part by the NSF Grant DMS 0700831. A part of his work has been done while visiting the Max Planck Institute in Bonn, Germany. He wishes to thank the institute for support.

Contents

1 Introduction 1

2 Expanding Random Maps 5
   2.1 Introductory Examples 5
   2.2 Preliminaries 8
   2.3 Expanding Random Maps 8
   2.4 Uniformly Expanding Random Maps 9
   2.5 Remarks on Expanding Random Mappings 10
   2.6 Visiting Sequences 11
   2.7 Spaces of Continuous and Hölder Functions 12
   2.8 Transfer Operator 13
   2.9 Distortion Properties 14

3 The RPF-Theorem 17
   3.1 Formulation of the Theorems 17
   3.2 Frequently used Auxiliary Measurable Functions 19
   3.3 Transfer Dual Operators 19
   3.4 Invariant Density 22
   3.5 Levels of Positive Cones of Hölder Functions 24
   3.6 Exponential Convergence of Transfer Operators 27
   3.7 Exponential Decay of Correlations 31
   3.8 Uniqueness 32
   3.9 Pressure Function 33
   3.10 Gibbs Property 35
   3.11 Some Comments on Uniformly Expanding
       Random Maps 37

4 Measurability, Pressure and Gibbs Condition 39
   4.1 Measurable Expanding Random Maps 39
   4.2 Measurability 41
   4.3 The Expected Pressure 42

    4.4   Ergodicity of μ 43

    4.5   Random Compact Subsets of Polish Spaces 43

5 Fractal Structure of Conformal Expanding Random Repellers 47

    5.1   Bowen's Formula 47

    5.2   Quasi-Deterministic and Essential Systems 51

    5.3   Random Cantor Set 54

6 Multifractal Analysis 57

    6.1   Concave Legendre Transform 57

    6.2   Multifractal Spectrum 59

    6.3   Analyticity of the Multifractal Spectrum
          for Uniformly Expanding Random Maps 67

7 Expanding in the Mean 69

    7.1   Definition of Maps Expanding in the Mean 69

    7.2   Associated Induced Map 70

    7.3   Back to the Original System 72

    7.4   An Example 73

8 Classical Expanding Random Systems 75

    8.1   Definition of Classical Expanding Random Systems 75

    8.2   Classical Conformal Expanding Random Systems 80

    8.3   Complex Dynamics and Brück and Büger Polynomial Systems 81

    8.4   Denker–Gordin Systems 84

    8.5   Conformal DG*-Systems 87

    8.6   Random Expanding Maps on Smooth Manifold 89

    8.7   Topological Exactness 89

    8.8   Stationary Measures 90

9 Real Analyticity of Pressure 93

    9.1   The Pressure as a Function of a Parameter 93

    9.2   Real Cones 97

    9.3   Canonical Complexification 100

    9.4   The Pressure is Real-Analytic 103

    9.5   Derivative of the Pressure 106

References 109

Index 111

Chapter 1
Introduction

In this monograph we develop the thermodynamical formalism for *measurable expanding random mappings*. This theory is then applied in the context of conformal expanding random mappings where we deal with the fractal geometry of fibers.

Distance expanding maps have been introduced for the first time in Ruelle's monograph [25]. A systematic account of the dynamics of such maps, including the thermodynamical formalism and the multifractal analysis, can be found in [24]. One of the main features of this class of maps is that their definition does not require any differentiability or smoothness condition. Distance expanding maps comprise symbol systems and expanding maps of smooth manifolds but go far beyond. This is also a characteristic feature of our approach.

We first define measurable expanding random maps. The randomness is modeled by an invertible ergodic transformation θ of a probability space $(X, \mathcal{B}, m)$. We investigate the dynamics of compositions

$$T_x^n = T_{\theta^{n-1}(x)} \circ \ldots \circ T_x, \quad n \geq 1,$$

where the $T_x : \mathcal{J}_x \to \mathcal{J}_{\theta(x)}$ $(x \in X)$ is a distance expanding mapping. These maps are only supposed to be measurably expanding in the sense that their expanding constant is measurable and a.e. $\gamma_x > 1$ or $\int \log \gamma_x \, dm(x) > 0$.

In so general setting we build the thermodynamical formalism for arbitrary Hölder continuous potentials φ_x. We show, in particular, the existence, uniqueness and ergodicity of a family of *Gibbs measures* $\{\nu_x\}_{x \in X}$. Following ideas of Kifer [17], these measures are first produced in a pointwise manner and then we carefully check their measurability. Often in the literature all fibers are contained in one and the same compact metric space and symbolic dynamics plays a prominent role. Our approach does not require the fibers to be contained in one metric space neither we need any Markov partitions or, even auxiliary, symbol dynamics.

Our results contain those in [5] and in [17] (see also the expository article [20]). Throughout the entire monograph where it is possible we avoid, in hypotheses, absolute constants. Our feeling is that in the context of random systems all (or at least

V. Mayer et al., *Distance Expanding Random Mappings, Thermodynamical Formalism, Gibbs Measures and Fractal Geometry*, Lecture Notes in Mathematics 2036, DOI 10.1007/978-3-642-23650-1_1, © Springer-Verlag Berlin Heidelberg 2011

as many as possible) absolute constants appearing in deterministic systems should become measurable functions. With this respect the thermodynamical formalism developed in here represents also, up to our knowledge, new achievements in the theory of random symbol dynamics or smooth expanding random maps acting on Riemannian manifolds.

Unlike recent trends aiming to employ the method of Hilbert metric (as for example in [12, 19, 26, 27]) our approach to the thermodynamical formalism stems primarily from the classical method presented by Bowen in [7] and undertaken by Kifer [17]. Developing it in the context of random dynamical systems we demonstrate that it works well and does not lead to too complicated (at least to our taste) technicalities. The measurability issue mentioned above results from convergence of the Perron–Frobenius operators. We show that this convergence is exponential, which implies exponential decay of correlations. These results precede investigations of a pressure function $x \mapsto P_x(\varphi)$ which satisfies the property

$$v_{\theta(x)}(T_x(A)) = e^{P_x(\varphi)} \int_A e^{-\varphi_x} dv_x,$$

where A is any measurable set such that $T_x|_A$ is injective. The integral, against the measure m on the base X, of this function is a central parameter $\mathscr{E}P(\varphi)$ of random systems called the *expected pressure*. If the potential φ depends analytically on parameters, we show that the expected pressure also behaves real analytically. We would like to mention that, contrary to the deterministic case, the spectral gap methods do not work in the random setting. Our proof utilizes the concept of complex cones introduced by Rugh in [26], and this is the only place, where we use the projective metric.

We then apply the above results mainly to investigate fractal properties of fibers of *conformal random systems*. They include Hausdorff dimension, Hausdorff and packing measures, as well as multifractal analysis. First, we establish a version of Bowen's formula (obtained in a somewhat different context in [6]) showing that the Hausdorff dimension of almost every fiber $\mathscr{I}_x$ is equal to h, the only zero of the expected pressure $\mathscr{E}P(\varphi_t)$, where $\varphi_t = -t \log |f'|$ and $t \in \mathbb{R}$. Then we analyze the behavior of h-dimensional Hausdorff and packing measures. It turned out that the random dynamical systems split into two categories. Systems from the first category, rather exceptional, behave like deterministic systems. We call them, therefore, *quasi-deterministic*. For them the Hausdorff and packing measures are finite and positive. Other systems, called *essentially random*, are rather generic. For them the h-dimensional Hausdorff measure vanishes while the h-packing measure is infinite. This, in particular, refutes the conjecture stated by Bogenschütz and Ochs in [6] that the h-dimensional Hausdorff measure of fibers is always positive and finite. In fact, the distinction between the quasi-deterministic and the essentially random systems is determined by the behavior of the Birkhoff sums

$$P_x^n(\varphi) = P_{\theta^{n-1}(x)}(\varphi) + \ldots + P_x(\varphi)$$

of the pressure function for potential $\varphi_h = -h \log |f'|$. If these sums stay bounded then we are in the quasi-deterministic case. On the other hand, if these sums are neither bounded below nor above, the system is called essentially random. The behavior of P_x^n, being random variables defined on X, the base map for our skew product map, is often governed by stochastic theorems such as the law of the iterated logarithm whenever it holds. This is the case for our primary examples, namely conformal DG-systems and classical conformal random systems. We are then in position to state that the quasi-deterministic systems correspond to rather exceptional case where the asymptotic variance $\sigma^2 = 0$. Otherwise the system is essential.

The fact that Hausdorff measures in the Hausdorff dimension vanish has further striking geometric consequences. Namely, almost all fibers of an essential conformal random system are not bi-Lipschitz equivalent to any fiber of any quasi-deterministic or deterministic conformal expanding system. In consequence almost every fiber of an essentially random system is not a geometric circle nor even a piecewise analytic curve. We then show that these results do hold for many explicit random dynamical systems, such as conformal DG-systems, classical conformal random systems, and, perhaps most importantly, Brück and Büger polynomial systems. As a consequence of the techniques we have developed, we positively answer the question of Brück and Büger (see [9] and Question 5.4 in [8]) of whether the Hausdorff dimension of almost all naturally defined random Julia set is strictly larger than 1. We also show that in this same setting the Hausdorff dimension of almost all Julia sets is strictly less than 2.

Concerning the multifractal spectrum of Gibbs measures on fibers, we show that the multifractal formalism is valid, i.e. the multifractal spectrum is Legendre conjugated to a temperature function. As usual, the temperature function is implicitly given in terms of the expected pressure. Here, the most important, although perhaps not most strikingly visible, issue is to make sure that there exists a set X_{ma} of full measure in the base such that the multifractal formalism works for all $x \in X_{ma}$.

If the system is in addition uniformly expanding then we provide real analyticity of the pressure function. This part is based on work by Rugh [27] and it is the only place where we work with the Hilbert metric. As a consequence and via Legendre transformation we obtain real analyticity of the multifractal spectrum.

Random transformations have already a long history and the present manuscript does, by no means, cover all its topics. Some of them can be found in Arnold's book [1] and in Kifer and Liu's chapter in [20]. Let us however mention some interesting results. Denote by $\mathcal{M}_m^1(T)$ the set of T-invariant measures from $\mathcal{M}_m^1(\mathcal{J})$. Let $\mu \in \mathcal{M}_m^1(T)$. The *fiber entropy* $h_\mu^r(T)$ of μ is given as follows. If $\mathcal{R} = \{R^1, R^2, \ldots, R^n\}$ is a finite partition of $\mathcal{J}$, then by $\mathcal{R}_x$ we denote the partition of $\mathcal{J}$ given by sets $R_x^k := R^k \cap \mathcal{J}_x$, $k = 1, \ldots, n$. Then

$$h_\mu^r(T) := \sup_{\mathcal{R}} \lim_{n \to \infty} \frac{1}{n} H_{\mu_x} \left(\bigvee_{i=0}^{n-1} \mathcal{R}_{\theta^i(x)} \right).$$

In fairly general random setting one can prove that this limit m-almost surely exists (see, e.g. [4]). Moreover, there is a following relation between the fiber entropy and the topological pressure called *Variational Principle* (see, e.g. [2, 4, 14])

$$\mathscr{E}P(\varphi) := \sup_{\mu \in (T)} \left(\int \varphi d\mu + h_\mu^r(T) \right).$$

It is also worth noting that in many cases the entropy and averaged positive Lyapunov exponents can satisfy so called Margulis–Ruelle inequality (see, e.g. [3]) or Pesin formula (see, e.g. [21]). We also refer the reader to [22].

We would like to thank Yuri I. Kifer for his remarks which improved the final version of this monograph.

Chapter 2
Expanding Random Maps

For the convenience of the reader, we first give some introductory examples. In the remaining part of this chapter we present the general framework of *expanding random maps*.

2.1 Introductory Examples

Before giving the formal definitions of expanding random maps, let us now consider some typical examples.

The first one is a known random version of the Sierpiński gasket (see, for example [15]). Let $\Delta = \Delta(A, B, C)$ be a triangle with vertexes A, B, C and choose $a \in (A, B)$, $b \in (B, C)$ and $c \in (C, A)$. Then we can associate to $x = (a, b, c)$ a map

$$f_x : \Delta(A, a, c) \cup \Delta(a, B, b) \cup \Delta(b, C, a) \to \Delta,$$

such that the restriction of f_x to each one of the three subtriangles is a affine map onto Δ. The map f_x is nothing else than the generator of a deterministic Sierpiński gasket. Note that this map can be made continuous by identifying the vertexes A, B, C (Fig. 2.1).

Now, suppose $x_1 = (a_1, b_1, c_1)$, $x_2 = (a_2, b_2, c_2)$, ... are chosen randomly which, for example, may mean that they form sequences of three dimensional independent and identically distributed (i.i.d.) random variables. Then they generate compact sets

$$\mathscr{J}_{x_1, x_2, x_3, \dots} = \bigcap_{n \geq 1} (f_{x_n} \circ \dots \circ f_{x_1})^{-1}(\Delta)$$

called *random Sierpiński gaskets* having the invariance property $f_{x_1}^{-1}(\mathscr{J}_{x_2, x_3, \dots}) = \mathscr{J}_{x_1, x_2, x_3, \dots}$. For a little bit simpler example of random Cantor sets we refer the reader to Sect. 5.3. In that example we provide a more detailed analysis of such random sets.

V. Mayer et al., *Distance Expanding Random Mappings, Thermodynamical Formalism,*   5
Gibbs Measures and Fractal Geometry, Lecture Notes in Mathematics 2036,
DOI 10.1007/978-3-642-23650-1_2, © Springer-Verlag Berlin Heidelberg 2011

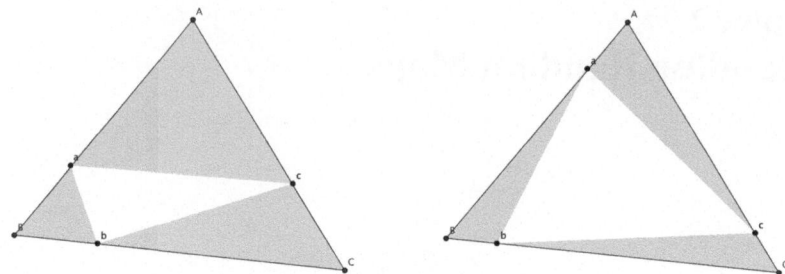

Fig. 2.1 Two different generators of Sierpiński gaskets

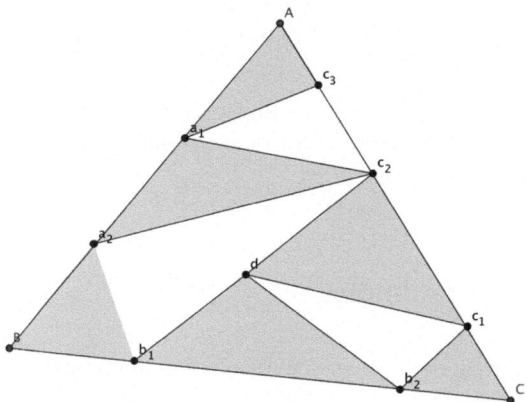

Fig. 2.2 A generator of degree 6

Such examples admit far going generalizations. First of all, we will consider much more general random choices than i.i.d. ones. We model randomness by taking a probability space $(X, \mathscr{B}, m)$ along with an invariant ergodic transformation θ : $X \to X$. This point of view was up to our knowledge introduced by the Bremen group (see [1]).

Another point is that the maps f_x that generate the random Sierpiński gasket have degree 3. In the sequel of this manuscript, we will allow the degree d_x of all maps to be different (see Fig. 2.2) and only require that the function $x \mapsto \log(d_x)$ is measurable.

Finally, the above examples are all expanding with an expanding constant

$$\gamma_x \geq \gamma > 1 \ .$$

As already explained in the introduction, the present monograph concerns random maps for which the expanding constants γ_x can be arbitrarily close to one. Furthermore, using an inducing procedure, we will even weaken this to the maps that are only expanding in the mean (see Chap. 7).

The example of random Sierpiński gasket is not conformal. Random iterations of rational functions or of holomorphic repellers are typical examples of conformal random dynamical systems. Random iterations of the quadratic family $f_c(z) = z^2 + c$ have been considered, for example, by Brück and Büger among others (see [8] and [9]). In this case, one chooses randomly a sequence of bounded parameters $c = (c_1, c_2, \ldots)$ and considers the dynamics of the family

$$F_{c_1,\ldots,c_n} = f_{c_n} \circ f_{c_{n_1}} \circ \ldots \circ f_{c_1}, \quad n \geq 1.$$

This leads to the dynamical invariant sets

$$\mathcal{K}_c = \{z \in \mathbb{C}; \ F_{c_1,\ldots,c_n}(z) \nrightarrow \infty\} \quad \text{and} \quad \mathcal{J}_c = \partial \mathcal{K}_c.$$

The set $\mathcal{K}_c$ is the filled in Julia set and $\mathcal{J}_c$ the Julia set associated to the sequence c.

The simplest case is certainly the one when we consider just two polynomials $z \mapsto z^2 + \lambda_1$ and $z \mapsto z^2 + \lambda_2$ and we build a random sequence out of them. Julia sets that come out of such a choice are presented in Fig. 2.3. Such random Julia sets are different objects as compared to the Julia sets for deterministic iteration of quadratic polynomials. But not only the pictures are different and intriguing, we

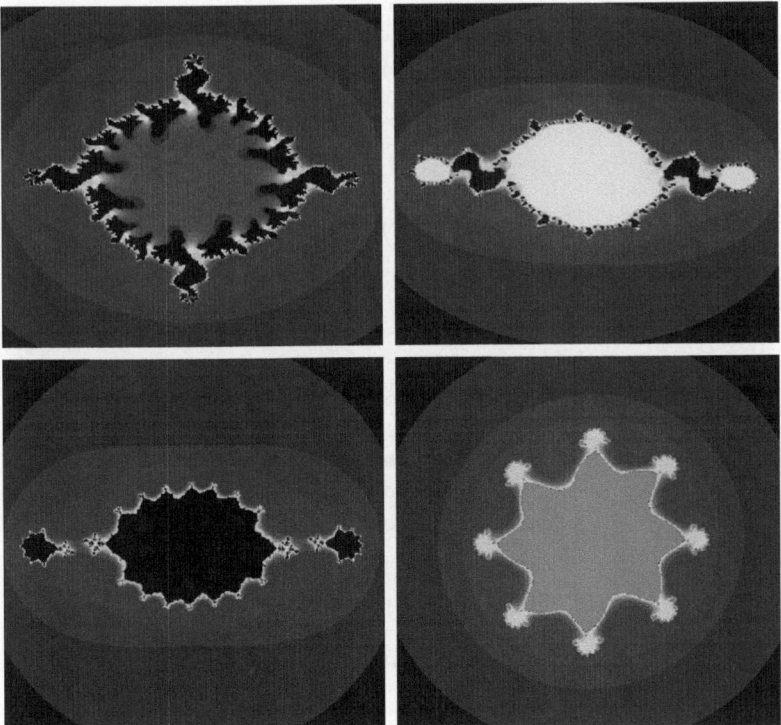

Fig. 2.3 Some quadratic random Julia sets

will see in Chap. 5 that also generically the fractal properties of such Julia sets are fairly different as compared with the deterministic case even if the dynamics are uniformly expanding. In Chap. 8 we present a more general class of examples and we explain their dynamical and fractal features.

2.2 Preliminaries

Suppose $(X, \mathscr{B}, m, \theta)$ is a measure preserving dynamical system with invertible and ergodic map $\theta : X \to X$ which is referred to as *the base map*. Assume further that $(\mathscr{J}_x, \rho_x)$, $x \in X$, are compact metric spaces normalized in size by $\mathrm{diam}_{\rho_x}(\mathscr{J}_x) \leq 1$. Let

$$\mathscr{J} = \bigcup_{x \in X} \{x\} \times \mathscr{J}_x. \tag{2.1}$$

We will denote by $B_x(z, r)$ the ball in the space $(\mathscr{J}_x, \varrho_x)$ centered at $z \in \mathscr{J}_x$ and with radius r. Frequently, for ease of notation, we will write $B(y, r)$ for $B_x(z, r)$, where $y = (x, z)$. Let

$$T_x : \mathscr{J}_x \to \mathscr{J}_{\theta(x)}, \quad x \in X,$$

be continuous mappings and let $T : \mathscr{J} \to \mathscr{J}$ be the associated skew-product defined by

$$T(x, z) = (\theta(x), T_x(z)). \tag{2.2}$$

For every $n \geq 0$ we denote $T_x^n := T_{\theta^{n-1}(x)} \circ \dots \circ T_x : \mathscr{J}_x \to \mathscr{J}_{\theta^n(x)}$. With this notation one has $T^n(x, y) = (\theta^n(x), T_x^n(y))$. We will frequently use the notation

$$x_n = \theta^n(x), \quad n \in \mathbb{Z}.$$

If it does not lead to misunderstanding we will identify $\mathscr{J}_x$ and $\{x\} \times \mathscr{J}_x$.

2.3 Expanding Random Maps

A map $T : \mathscr{J} \to \mathscr{J}$ is called a *expanding random map* if the mappings $T_x : \mathscr{J}_x \to \mathscr{J}_{\theta(x)}$ are continuous, open, and surjective, and if there exist a function $\eta : X \to \mathbb{R}_+$, $x \mapsto \eta_x$, and a real number $\xi > 0$ such that following conditions hold.

Uniform Openness. $T_x(B_x(z, \eta_x)) \supset B_{\theta(x)}(T_x(z), \xi)$ for every $(x, z) \in \mathscr{J}$.

Measurably Expanding. There exists a measurable function $\gamma : X \to (1, +\infty)$, $x \mapsto \gamma_x$ such that, for m-a.e. $x \in X$,

$$\varrho_{\theta(x)}(T_x(z_1), T_x(z_2)) \geq \gamma_x \varrho_x(z_1, z_2) \quad \text{whenever} \quad \varrho(z_1, z_2) < \eta_x, z_1, z_2 \in \mathscr{J}_x .$$

Measurability of the Degree. The map $x \mapsto \deg(T_x) := \sup_{y \in \mathscr{J}_{\theta(x)}} \# T_x^{-1}(\{y\})$ is measurable.

Topological Exactness. There exists a measurable function $x \mapsto n_\xi(x)$ such that

$$T_x^{n_\xi(x)}(B_x(z, \xi)) = \mathscr{J}_{\theta^{n_\xi(x)}(x)} \quad \text{for every } z \in \mathscr{J}_x \text{ and a.e. } x \in X. \quad (2.3)$$

Note that the measurably expanding condition implies that $T_x|_{B(z, \eta_x)}$ is injective for every $(x, z) \in \mathscr{J}$. Together with the compactness of the spaces $\mathscr{J}_x$ it yields the numbers $\deg(T_x)$ to be finite. Therefore the supremum in the condition of measurability of the degree is in fact a maximum.

In this work we consider two other classes of random maps. The first one consists of the *uniform expanding* maps defined below. These are expanding random maps with uniform control of measurable "constants". The other class we consider is composed of maps that are only *expanding in the mean*. These maps are defined like the expanding random maps above excepted that the uniform openness and the measurable expanding conditions are replaced by the following weaker conditions (see Chap. 7 for detailed definition).

1. All local inverse branches do exist.
2. The function γ in the measurable expanding condition is allowed to have values in $(0, \infty)$ but subjects only the condition

$$\int_X \log \gamma_x \, dm > 0.$$

We employ an inducing procedure to expanding in the mean random maps in order to reduce then to the case of random expanding maps. This is the content of Chap. 7 and the conclusion is that all the results of the present work valid for expanding random maps do also hold for expanding in the mean random maps.

2.4 Uniformly Expanding Random Maps

Most of this paper and, in particular, the whole thermodynamical formalism is devoted to measurable expanding systems. The study of fractal and geometric properties (which starts with Chap. 5), somewhat against our general philosophy, but with agreement with the existing tradition (see for example [5, 12, 17]), we will work mostly with *uniform* and *conformal* systems (the later are introduced in Chap. 5).

A expanding random map $T : \mathscr{J} \to \mathscr{J}$ is called *uniformly expanding* if

 - $\gamma_* := \inf_{x \in X} \gamma_x > 1$,
 - $\deg(T) := \sup_{x \in X} \deg(T_x) < \infty$,
 - $n_{\xi*} := \sup_{x \in X} n_\xi(x) < \infty$.

2.5 Remarks on Expanding Random Mappings

The conditions of uniform openness and measurably expanding imply that, for every $y = (x, z) \in \mathscr{J}$ there exists a unique continuous inverse branch

$$T_y^{-1} : B_{\theta(x)}(T(y), \xi) \to B_x(y, \eta_x)$$

of T_x sending $T_x(z)$ to z. By the measurably expanding property we have

$$\varrho(T_y^{-1}(z_1), T_y^{-1}(z_2)) \leq \gamma_x^{-1} \varrho(z_1, z_2) \quad \text{for} \quad z_1, z_2 \in B_{\theta(x)}(T(y), \xi) \qquad (2.4)$$

and

$$T_y^{-1}(B_{\theta(x)}(T(y), \xi)) \subset B_x(y, \gamma_x^{-1} \xi) \subset B_x(y, \xi).$$

Hence, for every $n \geq 0$, the composition

$$T_y^{-n} = T_y^{-1} \circ T_{T(y)}^{-1} \circ \ldots \circ T_{T^{n-1}(y)}^{-1} : B_{\theta^n(x)}(T^n(y), \xi) \to B_x(y, \xi) \qquad (2.5)$$

is well defined and has the following properties:

$$T_y^{-n} : B_{\theta^n(x)}(T^n(y), \xi) \to B_x(y, \xi)$$

is continuous,

$$T^n \circ T_y^{-n} = \mathrm{Id}|_{B_{\theta^n(x)}(T^n(y), \xi)}, \quad T_y^{-n}(T_x^n(z)) = z$$

and, for every $z_1, z_2 \in B_{\theta^n(x)}(T^n(y), \xi)$,

$$\varrho(T_y^{-n}(z_1), T_y^{-n}(z_2)) \leq (\gamma_x^n)^{-1} \varrho(z_1, z_2), \qquad (2.6)$$

where $\gamma_x^n = \gamma_x \gamma_{\theta(x)} \cdots \gamma_{\theta^{n-1}(x)}$. Moreover,

$$T_x^{-n}(B_{\theta^n(x)}(T^n(y), \xi)) \subset B_x(y, (\gamma_x^n)^{-1} \xi) \subset B_x(y, \xi). \qquad (2.7)$$

Lemma 2.1 *For every $r > 0$, there exists a measurable function $x \mapsto n_r(x)$ such that a.e.*

$$T_x^{n_r(x)}(B_x(z, r)) = \mathscr{J}_{\theta^{n_r(x)}(x)} \quad \text{for every } z \in \mathscr{J}_x. \qquad (2.8)$$

Moreover, there exists a measurable function $j : X \to \mathbb{N}$ such that a.e. we have

$$T_{x-j(x)}^{j(x)}(B_{x-j(x)}(z, \xi)) = \mathscr{J}_x \quad \text{for every } z \in \mathscr{J}_{x-j(x)}. \qquad (2.9)$$

Proof. In order to prove the first statement, consider $\gamma_0 > 1$ and let F be the set of $x \in X$ such that $\gamma_x \geq \gamma_0$. If γ_0 is sufficiently close to 1, then $m(F) > 0$. In the following section such a set will be called essential. In that section we also associate to such an essential set a set X'_{+F} (see (2.10)). Then for $x \in X'_{+F}$, the limit $\lim_{n \to \infty} (\gamma_x^n)^{-1} = 0$. Define

$$X_{+F,k} := \{x \in X'_{+F} : (\gamma_x^k)^{-1} \xi < r\}.$$

Then $X_{+F,k} \subset X_{+F,k+1}$ and $\bigcup_{k \in \mathbb{N}} X_{+F,k} = X'_{+F}$. By measurability of $x \mapsto \gamma_x$, $X_{+F,k}$ is a measurable set. Hence the function

$$X'_{+F} \ni x \mapsto n_r(x) := \min\{k : x \in X_{+F,k}\} + n_\xi(x)$$

is finite and measurable. By (2.7) and (2.3),

$$T_x^{n_r(x)}(B_x(z,r)) = \mathcal{J}_{\theta^{n_r(x)}(x)}.$$

In order to prove the existence of a measurable function $j : X \to \mathbb{N}$ define measurable sets

$$X_{\xi,n} := \{x \in X : n_\xi(x) \le n\}, \ X'_{\xi,n} := \theta^n(X_{\xi,n}) \text{ and } X'_\xi = \bigcup_{n \in \mathbb{N}} X'_{\xi,n}.$$

Then the map

$$X'_\xi \ni x \mapsto j(x) := \min\{n \in \mathbb{N} : x \in X'_{\xi,n}\}$$

satisfies (2.9) for $x \in X'_\xi$. Since $m(\theta^n(X_{\xi,n})) = m(X_{\xi,n}) \nearrow 1$ as n tends to ∞ we have $m(X'_\xi) = 1$. $\square$

2.6   Visiting Sequences

Let $F \in \mathcal{F}$ be a set with a positive measure. Define the sets

$$V_{+F}(x) := \{n \in \mathbb{N} : \theta^n(x) \in F\} \quad \text{and} \quad V_{-F}(x) := \{n \in \mathbb{N} : \theta^{-n}(x) \in F\}.$$

The set $V_{+F}(x)$ is called *visiting sequence* (of F at x). Then the set $V_{-F}(x)$ is just a visiting sequence for θ^{-1} and we also call it *backward visiting sequence*. By $n_j(x)$ we denote the jth-visit in F at x. Since $m(F) > 0$, by Birkhoff's Ergodic Theorem we have that

$$m(X'_{+F}) = m(X'_{-F}) = 1,$$

where

$$X'_{+F} := \left\{x \in X : V_{+F}(x) \text{ is infinite and } \lim_{j \to \infty} \frac{n_{j+1}(x)}{n_j(x)} = 1\right\} \qquad (2.10)$$

and X'_{-F} is defined analogously. The sets X'_{+F} and X'_{-F} are respectively called *forward* and *backward visiting for F*.

Let $\Psi(x, n)$ be a formula which depends on $x \in X$ and $n \in \mathbb{N}$. We say that $\Psi(x, n)$ holds *in a visiting way*, if there exists F with $m(F) > 0$ such that, for m-a.e. $x \in X'_{+F}$ and all $j \in \mathbb{N}$, the formula $\Psi(\theta^{n_j}(x), n_j(x))$ holds, where $(n_j(x))_{j=0}^{\infty}$ is the visiting sequence of F at x. We also say that $\Psi(x, n)$ holds *in a exhaustively visiting way*, if there exists a family $F_k \in \mathscr{F}$ with $\lim_{k \to \infty} m(F_k) = 1$ such that, for all k, m-a.e. $x \in X'_{+F_k}$, and all $j \in \mathbb{N}$, the formula $\Psi(\theta^{n_j}(x), n_j(x))$ holds, where $(n_j(x))_{j=0}^{\infty}$ is the visiting sequence of F_k at x.

Now, let $f_n : X \to \mathbb{R}$ be a sequence of measurable functions. We write that

$$\operatorname*{s-lim}_{n \to \infty} f_n = f,$$

if that there exists a family $F_k \in \mathscr{F}$ with $\lim_{k \to \infty} m(F_n) = 1$ such that, for all k and m-a.e. $x \in X'_{+F_k}$ and all $j \in \mathbb{N}$,

$$\lim_{j \to \infty} f_{n_j}(x) = f(x),$$

where $(n_j)_{j=0}^{\infty}$ is the visiting sequence of F_k at x.

Suppose that $g_1, \ldots, g_k : X \to \mathbb{R}$ is a finite collection of measurable functions and let $b_1, \ldots, b_n$ be a collection of real numbers. Consider the set

$$F := \bigcap_{i=1}^{k} g_i^{-1}((-\infty, b_i]).$$

If $m(F) > 0$, then F is called *essential* for $g_1, \ldots, g_k$ with constants $b_1, \ldots, b_n$ (or just *essential*, if we do not want explicitly specify functions and numbers). Note that by measurability of the functions $g_1, \ldots, g_k$, for every $\varepsilon > 0$ we can always find finite numbers $b_1, \ldots, b_n$ such that the essential set F for $g_1, \ldots, g_k$ with constants $b_1, \ldots, b_n$ has the measure $m(F) \geq 1 - \varepsilon$.

2.7 Spaces of Continuous and Hölder Functions

We denote by $\mathscr{C}(\mathscr{J}_x)$ the space of continuous functions $g_x : \mathscr{J}_x \to \mathbb{R}$ and by $\mathscr{C}(\mathscr{J})$ the space of functions $g : \mathscr{J} \to \mathbb{R}$ such that, for a.e. $x \in X$, $x \mapsto g_x := g|_{\mathscr{J}_x} \in \mathscr{C}(\mathscr{J}_x)$. The set $\mathscr{C}(\mathscr{J})$ contains the subspaces $\mathscr{C}^0(\mathscr{J})$ of functions for which the function $x \mapsto \|g_x\|_\infty$ is measurable, and $\mathscr{C}^1(\mathscr{J})$ for which the integral

$$\|g\|_1 := \int_X \|g_x\|_\infty \, dm(x) < \infty.$$

Now, fix $\alpha \in (0, 1]$. By $\mathscr{H}^\alpha(\mathscr{J}_x)$ we denote the space of Hölder continuous functions on $\mathscr{J}_x$ with an exponent α. This means that $\varphi_x \in \mathscr{H}^\alpha(\mathscr{J}_x)$ if and only if $\varphi_x \in \mathscr{C}(\mathscr{J}_x)$ and $v(\varphi_x) < \infty$ where

$$v_\alpha(\varphi_x) := \inf\{H_x : |\varphi(z_1) - \varphi(z_2)| \leq H_x \varrho_x^\alpha(z_1, z_2)\}, \qquad (2.11)$$

where the infimum is taken over all $z_1, z_2 \in \mathscr{J}_x$ with $\varrho(z_1, z_2) \leq \eta$.

A function $\varphi \in \mathscr{C}^1(\mathscr{J})$ is called *Hölder continuous with an exponent* α provided that there exists a measurable function $H : X \to [1, +\infty)$, $x \mapsto H_x$, such that $\log H \in L^1(m)$ and such that $v_\alpha(\varphi_x) \leq H_x$ for a.e. $x \in X$. We denote the space of all Hölder functions with fixed α and H by $\mathscr{H}^\alpha(\mathscr{J}, H)$ and the space of all α-Hölder functions by $\mathscr{H}^\alpha(\mathscr{J}) = \bigcup_{H \geq 1} \mathscr{H}^\alpha(\mathscr{J}, H)$.

2.8   Transfer Operator

For every function $g : \mathscr{J} \to \mathbb{C}$ and a.e. $x \in X$ let

$$S_n g_x = \sum_{j=0}^{n-1} g_x \circ T_x^j, \qquad (2.12)$$

and, if $g : X \to \mathbb{C}$, then $S_n g = \sum_{j=0}^{n-1} g \circ \theta^j$. Let φ be a function in the Hölder space $\mathscr{H}^\alpha(\mathscr{J})$. For every $x \in X$, we consider the *transfer operator* $\mathscr{L}_x = \mathscr{L}_{\varphi, x} : \mathscr{C}(\mathscr{J}_x) \to \mathscr{C}(\mathscr{J}_{\theta(x)})$ given by the formula

$$\mathscr{L}_x g_x(w) = \sum_{T_x(z) = w} g_x(z) e^{\varphi_x(z)}, \quad w \in \mathscr{J}_{\theta(x)}. \qquad (2.13)$$

It is obviously a positive linear operator and it is bounded with the norm bounded above by

$$\|\mathscr{L}_x\|_\infty \leq \deg(T_x) \exp(\|\varphi\|_\infty). \qquad (2.14)$$

This family of operators gives rise to the global operator $\mathscr{L} : \mathscr{C}(\mathscr{J}) \to \mathscr{C}(J)$ defined as follows:

$$(\mathscr{L}g)_x(w) = \mathscr{L}_{\theta^{-1}(x)} g_{\theta^{-1}(x)}(w).$$

For every $n > 1$ and a.e. $x \in X$, we denote

$$\mathscr{L}_x^n := \mathscr{L}_{\theta^{n-1}(x)} \circ \ldots \circ \mathscr{L}_x : \mathscr{C}(\mathscr{J}_x) \to \mathscr{C}(\mathscr{J}_{\theta^n(x)}).$$

Note that

$$\mathscr{L}_x^n g_x(w) = \sum_{z \in T_x^{-n}(w)} g_x(z) e^{S_n \varphi_x(z)}, \quad w \in \mathscr{J}_{\theta^n(x)}, \qquad (2.15)$$

where $S_n \varphi_x(z)$ has been defined in (2.12). The dual operator $\mathscr{L}_x^*$ maps $C^*(\mathscr{J}_{\theta(x)})$ into $C^*(\mathscr{J}_x)$.

2.9   Distortion Properties

Lemma 2.2 *Let $\varphi \in \mathcal{H}^\alpha(\mathcal{J}, H)$, let $n \geq 1$ and let $y = (x,z) \in \mathcal{J}$. Then*

$$|S_n\varphi_x(T_y^{-n}(w_1)) - S_n\varphi_x(T_y^{-n}(w_2))| \leq \varrho^\alpha(w_1, w_2) \sum_{j=0}^{n-1} H_{\theta^j(x)}(\gamma_{\theta^j(x)}^{n-j})^{-\alpha}$$

for all $w_1, w_2 \in B(T_x^n(z), \xi)$.

Proof. We have by (2.6) and Hölder continuity of φ that

$$|S_n\varphi_x(T_y^{-n}(w_1)) - S_n\varphi_x(T_y^{-n}(w_2))| \leq \sum_{j=0}^{n-1} |\varphi_x(T_x^j(T_y^{-n}(w_1))) - \varphi_x(T_x^j(T_y^{-n}(w_2)))|$$

$$= \sum_{j=0}^{n-1} \left|\varphi_x(T_{T_x^j(y)}^{-(n-j)}(w_1)) - \varphi_x(T_{T_x^j(y)}^{-(n-j)}(w_2))\right|$$

$$\leq \sum_{j=0}^{n-1} \varrho^\alpha(T_{T_x^j(x)}^{-(n-j)}(w_1), T_{T_x^j(x)}^{-(n-j)}(w_2)) H_{\theta^j(x)},$$

hence $|S_n\varphi_x(T_y^{-n}(w_1)) - S_n\varphi_x(T_y^{-n}(w_2))| \leq \varrho^\alpha(w_1, w_2) \sum_{j=0}^{n-1} H_{\theta^j(x)}(\gamma_{\theta^j(x)}^{n-j})^{-\alpha}$. $\qquad\square$

Set

$$Q_x := Q_x(H) = \sum_{j=1}^{\infty} H_{\theta^{-j}(x)}(\gamma_{\theta^{-j}(x)}^j)^{-\alpha}. \tag{2.16}$$

Lemma 2.3 *The function $x \mapsto Q_x$ is measurable and m-a.e. finite. Moreover, for every $\varphi \in \mathcal{H}^\alpha(\mathcal{J}, H)$,*

$$|S_n\varphi_x(T_y^{-n}(w_1)) - S_n\varphi_x(T_y^{-n}(w_2))| \leq Q_{\theta^n(x)}\varrho^\alpha(w_1, w_2)$$

for all $n \geq 1$, a.e. $x \in X$, every $z \in \mathcal{J}_x$ and $w_1, w_2 \in B(T^n(z), \xi)$ and where again $y = (x,z)$.

Proof. The measurability of Q_x follows directly from (2.16). Because of Lemma 2.2 we are only left to show that Q_x is m-a.e. finite. Let χ be a positive real number less or equal to $\int \log \gamma_x dm(x)$. Then, using Birkhoff's Ergodic Theorem for θ^{-1}, we get that

$$\liminf_{j \to \infty} \frac{1}{j} \sum_{k=0}^{j-1} \log \gamma_{\theta^{-j}(x)} \geq \chi$$

for m-a.e. $x \in X$. Therefore, there exists a measurable function $C_\gamma : X \to [1, +\infty)$ m-a.e. finite such that $C_\gamma^{-1}(x)e^{j\chi/2} \leq \gamma_{\theta^{-j+1}(x)}^j$ for all $j \geq 0$ and a.e. $x \in X$.

Moreover, since $\log H \in L^1(m)$ it follows again from Birkhoff's Ergodic Theorem that

$$\lim_{j \to \infty} \frac{1}{j} \log H_{\theta^{-j}(x)} = 0 \quad m\text{-}a.e.$$

There thus exists a measurable function $C_H : X \to [1, +\infty)$ such that

$$H_{\theta^j(x)} \leq C_H(x)e^{j\alpha\chi/4} \quad \text{and} \quad H_{\theta^{-j}(x)} \leq C_H(x)e^{j\alpha\chi/4} \tag{2.17}$$

for all $j \geq 0$ and a.e. $x \in X$. Then, for a.e. $x \in X$, all $n \geq 0$ and all $a \geq j \geq n-1$, we have

$$H_{\theta^j(x)} = H_{\theta^{-(n-j)}(\theta^n(x))} \leq C_H(\theta^n(x))e^{(n-j)\alpha\chi/4}.$$

Therefore, still with $x_n = \theta^n(x)$,

$$Q_{x_n} = \sum_{j=0}^{n-1} H_{x_j}(\gamma_{x_j}^{n-j})^{-\alpha} \leq \sum_{j=0}^{n-1} C_H(x_n)e^{(n-j)\alpha\chi/4}C_\gamma^\alpha(x_{n-1})e^{-\alpha(n-j)\chi/2}$$

$$\leq C_\gamma^\alpha(x_{n-1})C_H(x_n) \sum_{j=0}^{n-1} e^{-\alpha(n-j)\chi/4} \leq C_\gamma^\alpha(x_{n-1})C_H(x_n)(1 - e^{-\alpha\chi/4})^{-1}.$$

Hence

$$Q_x \leq C_\gamma^\alpha(\theta^{-1}(x))C_H(x)(1 - e^{-\alpha\chi/4})^{-1} < +\infty.$$

$\square$

Chapter 3
The RPF-Theorem

We now establish a version of Ruelle–Perron–Frobenius (RPF) Theorem along with a mixing property. Notice that this quite substantial fact is proved without any measurable structure on the space $\mathscr{J}$. In particular, we do not address measurability issues of λ_x and q_x. In order to obtain this measurability we will need and we will impose a natural measurable structure on the space $\mathscr{J}$. This will be done in the next chapter.

3.1 Formulation of the Theorems

Let $T : \mathscr{J} \to \mathscr{J}$ be a expanding random map. Denote by $\mathcal{M}^1(\mathscr{J}_x)$ the set of all Borel probability measures on $\mathscr{J}_x$. A family of measures $\{\mu_x\}_{x \in X}$ such that $\mu_x \in \mathcal{M}^1(\mathscr{J}_x)$ is called T-invariant if $\mu_x \circ T_x^{-1} = \mu_{\theta(x)}$ for a.e. $x \in X$.

This chapter is devoted to the thermodynamical formalism. The main results proved here are listed below.

Theorem 3.1 *Let $\varphi \in \mathcal{H}^\alpha(\mathscr{J})$ and let $\mathscr{L} = \mathscr{L}_\varphi$ be the associated transfer operator. Then the following holds.*

1. There exists a unique family of probability measures $\nu_x \in \mathcal{M}(\mathscr{J}_x)$ such that

$$\mathscr{L}_x^* \nu_{\theta(x)} = \lambda_x \nu_x \quad \text{where} \quad \lambda_x = \nu_{\theta(x)}(\mathscr{L}_x \mathbb{1}) \quad \text{m-a.e.} \tag{3.1}$$

2. There exists a unique function $q \in \mathscr{C}^0(\mathscr{J})$ such that m-a.e.

$$\mathscr{L}_x q_x = \lambda_x q_{\theta(x)} \quad \text{and} \quad \nu_x(q_x) = 1. \tag{3.2}$$

Moreover, $q_x \in \mathcal{H}^\alpha(\mathscr{J}_x)$ for a.e. $x \in X$.

3. The family of measures $\{\mu_x := q_x \nu_x\}_{x \in X}$ is T-invariant.

V. Mayer et al., *Distance Expanding Random Mappings, Thermodynamical Formalism,*
Gibbs Measures and Fractal Geometry, Lecture Notes in Mathematics 2036,
DOI 10.1007/978-3-642-23650-1_3, © Springer-Verlag Berlin Heidelberg 2011

Theorem 3.2

1. Let

$$\hat{\varphi}_x = \varphi_x + \log q_x - \log q_{\theta(x)} \circ T - \log \lambda_x.$$

Denote $\hat{\mathcal{L}} := \mathcal{L}_{\hat{\varphi}}$. *Then, for a.e.* $x \in X$ *and all* $g_x \in C(\mathcal{J}_x)$,

$$\hat{\mathcal{L}}_x^n g_x \xrightarrow[n \to \infty]{} \int g_x q_x d\nu_x.$$

2. Let $\tilde{\varphi}_x = \varphi_x - \log \lambda_x$. *Denote* $\tilde{\mathcal{L}} := \mathcal{L}_{\tilde{\varphi}}$. *There exist a constant* $B < 1$ *and a measurable function* $A : X \to (0, \infty)$ *such that for every function* $g \in \mathcal{C}^0(\mathcal{J})$ *with* $g_x \in \mathcal{H}^\alpha(\mathcal{J}_x)$ *there exists a measurable function* $A_g : X \to (0, \infty)$ *for which*

$$\|(\tilde{\mathcal{L}}^n g)_x - \left(\int g_{\theta^{-n}(x)} d\nu_{\theta^{-n}(x)}\right) q_x \|_\infty \le A_g (\theta^{-n}(x)) A(x) B^n$$

for a.e. $x \in X$ *and every* $n \ge 1$.

3. There exists $B < 1$ *and a measurable function* $A' : X \to (0, \infty)$ *such that for every* $f_{\theta^n(x)} \in L^1(\mu_{\theta^n(x)})$ *and every* $g_x \in \mathcal{H}^\alpha(\mathcal{J}_x)$,

$$\left| \mu_x\left((f_{\theta^n(x)} \circ T_x^n) g_x\right) - \mu_{\theta^n(x)}(f_{\theta^n(x)}) \mu_x(g_x) \right|$$

$$\le \mu_{\theta^n(x)}(|f_{\theta^n(x)}|) A'(\theta^n(x)) \left(\int |g_x| d\mu_x + 4 \frac{v_\alpha(g_x q_x)}{Q_x} \right) B^n.$$

A collection of measures $\{\mu_x\}_{x \in X}$ such that $\mu_x \in \mathcal{M}^1(\mathcal{J}_x)$ is called a *Gibbs family* for $\varphi \in \mathcal{H}^\alpha(\mathcal{J})$ provided that there exists a measurable function $D_\varphi : X \to [1, +\infty)$ and a function $x \mapsto P_x$, called *the pseudo-pressure function*, such that

$$(D_\varphi(x) D_\varphi(\theta^n(x)))^{-1} \le \frac{\mu_x(T_y^{-n}(B(T^n(y), \xi)))}{\exp(S_n \varphi(y) - S_n P_x)} \le D_\varphi(x) D_\varphi(\theta^n(x)) \quad (3.3)$$

for every $n \ge 0$, a.e. $x \in X$ and every $z \in \mathcal{J}_x$ and with $y = (x, z)$.

Towards proving uniqueness type result for Gibbs families we introduce the following concept. Notice that in the case of random compact subsets of a Polish space (see Sect. 4.5) this condition always holds (see Lemma 4.11).

Measurability of Cardinality of Covers. There exists a measurable function $X \ni x \mapsto a_x \in \mathbb{N}$ such that for almost every $x \in X$ there exists a finite sequence $w_x^1, \ldots, w_x^{a_x} \in \mathcal{J}_x$ such that $\bigcup_{j=1}^{a_x} B(w_x^j, \xi) = \mathcal{J}_x$.

Theorem 3.3 *The collections* $\{\nu_x\}_{x \in X}$ *and* $\{\mu_x\}_{x \in X}$ *are Gibbs families. Moreover, if* $\mathcal{J}$ *satisfies the condition of measurability of cardinality of covers and if* $\{\nu'_x\}_{x \in X}$ *is a Gibbs family, then* ν'_x *and* ν_x *are equivalent for almost every* $x \in X$.

3.2 Frequently used Auxiliary Measurable Functions

Some technical measurable functions appear throughout the paper so frequently that, for convenience of the reader, we decided to collect them in this section together. However, the reader may skip this part now without any harm and come back to it when it is appropriately needed.

First, define

$$D_\xi(x) := \left(\deg T_x^n\right)^{-1} \exp(-2\|S_n\varphi_x\|_\infty) \tag{3.4}$$

with $n = n_\xi(x)$ being the index given by the topological exactness condition (cf. (2.3)). Then, let $j = j(x)$ be the number given by Lemma 2.1 and define

$$C_\varphi(x) := e^{Q_{x-j}} \deg(T_{x-j}^j) \max\left\{\exp(2\|S_k\varphi_{x-k}\|_\infty) : 0 \le k \le j\right\} \ge 1. \tag{3.5}$$

Now let $s > 1$. Put

$$C_{\min}(x) := e^{-sQ_x} e^{-\|S_j\varphi_{x-j}\|_\infty} \le 1 \tag{3.6}$$

and

$$C_{\max}(x) := e^{sQ_x} \deg\left(T_x^n\right) \exp(2\|S_n\varphi_x\|_\infty), \tag{3.7}$$

where $n := n_\xi(x)$. Then we define

$$\beta_x(s) := \frac{C_{\min}(x)}{C_\varphi(x)} \cdot \inf_{r \in (0,\xi]} \frac{1 - \exp\left(-(s-1)H_{x-1}\gamma_{x-1}^{-\alpha} r^\alpha\right)}{1 - \exp(-2sQ_x r^\alpha)}. \tag{3.8}$$

Since by (2.16)

$$sQ_x = sQ_{x-1}\gamma_{x-1}^{-\alpha} + sH_{x-1}\gamma_{x-1}^{-\alpha}, \tag{3.9}$$

$$(sQ_{x-1} + H_{x-1})\gamma_{x-1}^{-\alpha} = sQ_x - (s-1)H_{x-1}\gamma_{x-1}^{-\alpha}. \tag{3.10}$$

This, together with (3.5) and (3.6), gives us that

$$0 < \beta_x(s) = \frac{C_{\min}(x)}{C_\varphi(x)} \frac{(s-1)H_{x-1}\gamma_{x-1}^{-\alpha}}{2sQ_x} < \frac{C_{\min}(x)}{C_\varphi(x)} \le 1.$$

3.3 Transfer Dual Operators

In order to prove Theorem 3.1 we fix a point $x_0 \in X$ and, as the first step, we reduce the base space X to the orbit

$$\mathscr{O}_{x_0} = \{\theta^n(x_0), n \in \mathbb{Z}\}.$$

The motivation for this is that then we can deal with a sequentially topological compact space on which the transfer (or related) operators act continuously. Our conformal measure then can be produced, for example, by the methods of the fixed point theory, similarly as in the deterministic case.

Removing a set of measure zero, if necessary, we may assume that this orbit is chosen so that all the involved measurable functions are defined and finite on the points of $\mathcal{O}_{x_0}$. For every $x \in \mathcal{O}_{x_0}$, let $\varphi_x \in \mathscr{C}(\mathscr{J}_x)$ be the continuous potential of the transfer operator $\mathscr{L}_x : C(\mathscr{J}_x) \to C(\mathscr{J}_{\theta(x)})$ which has been defined in (2.13).

Proposition 3.4 *There exists probability measures $v_x \in M(\mathscr{J}_x)$ such that*

$$\mathscr{L}_x^* v_{\theta(x)} = \lambda_x v_x \quad \text{for every } x \in \mathcal{O}_{x_0},$$

where

$$\lambda_x := \mathscr{L}_x^*(v_{\theta(x)})(\mathbb{1}) = v_{\theta(x)}(\mathscr{L}_x \mathbb{1}). \tag{3.11}$$

Proof. Let $\mathscr{C}^*(\mathscr{J}_x)$ be the dual space of $\mathscr{C}(\mathscr{J}_x)$ equipped with the weak* topology. Consider the product space

$$\mathscr{D}(\mathcal{O}_{x_0}) := \prod_{x \in \mathcal{O}_{x_0}} \mathscr{C}^*(\mathscr{J}_x)$$

with the product topology. This is a locally convex topological space and the set

$$\mathscr{P}(\mathcal{O}_{x_0}) := \prod_{x \in \mathcal{O}_{x_0}} \mathscr{M}^1(\mathscr{J}_x)$$

is a compact subset of $\mathscr{D}(\mathcal{O}_{x_0})$. A simple observation is that the map

$$\Psi_x : \mathscr{M}^1(\mathscr{J}_{\theta(x)}) \to \mathscr{M}^1(\mathscr{J}_x)$$

defined by

$$\Psi_x(v_{\theta(x)}) = \frac{\mathscr{L}_x^* v_{\theta(x)}}{\mathscr{L}_x^* v_{\theta(x)}(\mathbb{1})}$$

is weakly continuous. Consider then the global map $\Psi : \mathscr{P}(\mathcal{O}_{x_0}) \to \mathscr{P}(\mathcal{O}_{x_0})$ given by

$$v = (v_x)_{x \in \mathcal{O}_{x_0}} \longmapsto \Psi(v) = \left(\Psi_x v_{\theta(x)}\right)_{x \in \mathcal{O}_{x_0}}.$$

Weak continuity of the Ψ_x implies continuity of Ψ with respect to the coordinate convergence. Since the space $\mathscr{P}(\mathcal{O}_{x_0})$ is a compact subset of a locally convex topological space, we can apply the Schauder–Tychonoff fixed point theorem to get $v \in \mathscr{P}(\mathcal{O}_{x_0})$ fixed point of Ψ, i.e.

$$\mathscr{L}_x^* v_{\theta(x)} = \lambda_x v_x \quad \text{where } \lambda_x = \mathscr{L}_x^* v_{\theta(x)}(\mathbb{1}) = v_{\theta(x)}(\mathscr{L}_x(\mathbb{1}))$$

for every $x \in \mathcal{O}_{x_0}$. $\square$

Remark 3.5 *The relation* (3.11) *implies*

$$\inf_{y \in \mathcal{J}_x} e^{\varphi_x(y)} \leq \lambda_x \leq \|\mathcal{L}_x \mathbb{1}\|_{\infty}. \tag{3.12}$$

A straightforward adaptation of the proof of Proposition 2.2 in [13] leads to the following, to Proposition 3.4 equivalent, characterization of Gibbs states: if $T^n_{x|A}$ is injective, then

$$\nu_{\theta^n(x)}(T^n_x(A)) = \lambda^n_x \int_A e^{-S_n \varphi} d\nu_x. \tag{3.13}$$

Here is one more useful estimate.

Lemma 3.6 *For every* $x \in \mathcal{O}_{x_0}$ *and* $n \geq 1$,

$$\inf_{z \in \mathcal{J}_x} \exp\left(S_n \varphi_x(z)\right) \leq \frac{\lambda^n_x}{\deg(T^n_x)} \leq \sup_{z \in \mathcal{J}_x} \exp\left(S_n \varphi_x(z)\right). \tag{3.14}$$

Moreover, for every $z \in \mathcal{J}_x$ *and every* $r > 0$,

$$\nu_x(B(z, r)) \geq D(x, r), \tag{3.15}$$

where

$$D(x, r) := \left(\deg(T^N_x)\right)^{-1} \inf_{z \in \mathcal{J}_x} \exp\left(\inf_{a \in B(z,r)} S_N \varphi_x(a) - \sup_{b \in B(z,r)} S_N \varphi_x(b)\right) \tag{3.16}$$

with $N = n_r(x)$ *being the index given by Lemma 2.1. It follows that the set* $\mathcal{J}_x$ *is a topological support of* ν_x. *In particular, with* $D_\xi(x)$ *defined in* (3.4),

$$\nu_x(B(z, \xi)) \geq D_\xi(x). \tag{3.17}$$

Proof. The inequalities (3.14) immediately follow from

$$\nu_{\theta^n(x)}(\mathcal{L}^n_x \mathbb{1}) = ((\mathcal{L}^n_x)^* \nu_{\theta^n(x)})(\mathbb{1}) = \lambda^n_x \nu_x(\mathbb{1}) = \lambda^n_x.$$

Now fix an arbitrary $z \in \mathcal{J}_x$ and $r > 0$. Put $n = n_r(x)$ (see Lemma 2.1). Then, by (3.13),

$$\nu_x(B(z, r)) \lambda^n_x \sup_{a \in B(z,r)} e^{-S_n \varphi_x(a)} \geq \lambda^n_x \int_{B(z,r)} e^{-S_n \varphi_x} d\nu_x \geq 1,$$

which implies (3.15). $\qquad\qquad\square$

3.4 Invariant Density

Consider now the normalized operator $\tilde{\mathscr{L}}$ given by

$$\tilde{\mathscr{L}}_x = \lambda_x^{-1} \mathscr{L}_x, \quad x \in X. \tag{3.18}$$

Proposition 3.7 *For every $x \in \mathcal{O}_{x_0}$, there exists a function $q_x \in \mathscr{H}^\alpha(\mathscr{J}_x)$ such that*

$$\tilde{\mathscr{L}}_x q_x = q_{\theta(x)} \quad and \quad \int_{\mathscr{J}_x} q_x \, dv_x = 1.$$

In addition,

$$q_x(z_1) \le \exp\{Q_x \varrho^\alpha(z_1, z_2)\} q_x(z_2)$$

for all $z_1, z_2 \in \mathscr{J}_x$ with $\varrho(z_1, z_2) \le \xi$, and

$$1/C_\varphi(x) \le q_x \le C_\varphi(x), \tag{3.19}$$

where C_φ was defined in (3.5).

In order to prove this statement we first need a good uniform distortion estimate.

Lemma 3.8 *For all $w_1, w_2 \in \mathscr{J}_x$ and $n \ge 1$*

$$\frac{\tilde{\mathscr{L}}_{x-n}^n \mathbb{1}(w_1)}{\tilde{\mathscr{L}}_{x-n}^n \mathbb{1}(w_2)} = \frac{\mathscr{L}_{x-n}^n \mathbb{1}(w_1)}{\mathscr{L}_{x-n}^n \mathbb{1}(w_2)} \le C_\varphi(x), \tag{3.20}$$

where C_φ is given by (3.5). If in addition $\varrho(w_1, w_2) \le \xi$, then

$$\frac{\tilde{\mathscr{L}}_{x-n}^n \mathbb{1}(w_1)}{\tilde{\mathscr{L}}_{x-n}^n \mathbb{1}(w_2)} \le \exp\{Q_x \varrho^\alpha(w_1, w_2)\}. \tag{3.21}$$

Moreover,

$$1/C_\varphi(x) \le \tilde{\mathscr{L}}_{x-n}^n \mathbb{1}(w) \le C_\varphi(x) \quad for\ every\ w \in \mathscr{J}_x\ and\ n \ge 1. \tag{3.22}$$

Proof. First, (3.21) immediately follows from Lemma 2.3. Notice also that

$$\exp\left(Q_x \varrho^\alpha(w_1, w_2)\right) \le \exp Q_x \tag{3.23}$$

since $\mathrm{diam}(\mathscr{J}_x) \le 1$. The global version of (3.20) can be proved as follows. If $n = 0, \ldots, j(x)$, then for every $w_1, w_2 \in \mathscr{J}_x$,

$$\mathscr{L}_{x-n}^n \mathbb{1}(w_1) \le \frac{\deg(T_{x-n}^n) \exp(\|S_n \varphi_{x-n}\|_\infty)}{\exp(-\|S_n \varphi_{x-n}\|_\infty)} \mathscr{L}_{x-n}^n \mathbb{1}(w_2) \le C_\varphi(x) \mathscr{L}_{x-n}^n \mathbb{1}(w_2).$$

Next, let $n > j := j(x)$. Take $w'_1 \in T_{x-j}^{-j}(w_1)$ such that

$$e^{S_j \varphi(w'_1)} \mathscr{L}_{x-n}^{n-j} \mathbb{1}(w'_1) = \sup_{y \in T_{x-j}^{-j}(w_1)} \left(e^{S_j \varphi(y)} \mathscr{L}_{x-n}^{n-j} \mathbb{1}(y) \right)$$

and $w'_2 \in T_{x-j}^{-j}(w_2)$ such that $\varrho_{x-j}(w'_1, w'_2) \le \xi$. Then, by (3.21) and (3.23),

$$\mathscr{L}_{x-n}^n \mathbb{1}(w_1) = \mathscr{L}_{x-j}^j (\mathscr{L}_{x-n}^{n-j} \mathbb{1})(w_1) \le \deg(T_{x-j}^j) e^{S_j \varphi(w'_1)} \mathscr{L}_{x-n}^{n-j} \mathbb{1}(w'_1)$$

$$\le \deg(T_{x-j}^j) e^{S_j \varphi(w'_1)} e^{Q_{x-j}} \mathscr{L}_{x-n}^{n-j} \mathbb{1}(w'_2) \le C_\varphi(x) \mathscr{L}_{x-n}^n \mathbb{1}(w_2).$$

This shows (3.20). By Proposition 3.4

$$\int_{\mathscr{J}_x} \tilde{\mathscr{L}}_{x-n}^n (\mathbb{1}) dv_x = \int_{\mathscr{J}_{x-n}} \mathbb{1} dv_{x-n} = 1, \tag{3.24}$$

which implies the existence of $w, w' \in \mathscr{J}_x$ such that $\tilde{\mathscr{L}}_{x-n}^n \mathbb{1}(w) \le 1$ and $\tilde{\mathscr{L}}_{x-n}^n \mathbb{1}(w') \ge 1$. Therefore, by the already proved part of this lemma, we get (3.22). $\square$

Proof. [Proof of Proposition 3.7] Let $x \in \mathscr{O}_{x_0}$. Then by Lemma 3.8, for every $k \ge 0$ and all $w_1, w_2 \in \mathscr{J}_x$ with $\varrho(w_1, w_2) \le \xi$, we have that

$$|\tilde{\mathscr{L}}_{x-k}^k \mathbb{1}(w_1) - \tilde{\mathscr{L}}_{x-k}^k \mathbb{1}(w_2)| \le C_\varphi(x) 2 Q_x \varrho^\alpha(w_1, w_2)$$

and $1/C_\varphi(x) \le \tilde{\mathscr{L}}_{x-k}^k \mathbb{1} \le C_\varphi(x)$. It follows that the sequence

$$q_{x,n} := \frac{1}{n} \sum_{k=0}^{n-1} \tilde{\mathscr{L}}_{x-k}^k \mathbb{1}, \quad n \ge 1$$

is equicontinuous for every $x \in \mathscr{O}_{x_0}$. Therefore, there exists a sequence $n_j \to \infty$ such that $q_{x,n_j} \to q_x$ uniformly for every x of the countable set $\mathscr{O}_{x_0}$. The functions q_x have all the required properties. $\square$

Let

$$\mu_x := q_x v_x, \tag{3.25}$$

and let $\hat{\mathscr{L}}_x := \mathscr{L}_{\hat{\varphi}, x}$ be the transfer operator with potential

$$\hat{\varphi}_x = \varphi_x + \log q_x - \log q_{\theta(x)} \circ T_x - \log \lambda_x.$$

Then

$$\hat{\mathscr{L}}_x g_x = \frac{1}{q_{\theta(x)}} \tilde{\mathscr{L}}_x (g_x q_x) \quad \text{for every } g_x \in L^1(\mu_x). \tag{3.26}$$

Consequently

$$\mathscr{L}_x \mathbb{1}_x = \mathbb{1}_{\theta(x)}. \tag{3.27}$$

Lemma 3.9 *For all* $g_{\theta(x)} \in L^1(\mu_{\theta(x)}) = L^1(\nu_{\theta(x)})$, *we have*

$$\mu_x(g_{\theta(x)} \circ T_x) = \mu_{\theta(x)}(g_{\theta(x)}). \tag{3.28}$$

Proof. From conformality of ν_x (see Proposition 3.4) it follows that

$$\hat{\mathscr{L}}_x^*(\mu_{\theta(x)})(g_x) = \int_{\mathscr{J}_{\theta(x)}} \hat{\mathscr{L}}_x(g_x) d\mu_{\theta(x)} = \lambda_x^{-1} \int_{\mathscr{J}_{\theta(x)}} (\mathscr{L}_x g_x q_x) d\nu_{\theta(x)}$$

$$= \lambda_x^{-1} \hat{\mathscr{L}}_x^*(\nu_{\theta(x)})(g_x q_x) = \nu_x(g_x q_x) = \mu_x(g_x). \tag{3.29}$$

So, if $f_x \cdot (g_{\theta(x)} \circ T_x) \in L^1(\mu_x)$, then

$$\mu_x\big((g_{\theta(x)} \circ T_x) f_x\big) = \hat{\mathscr{L}}_x^*(\mu_{\theta(x)})\big((g_{\theta(x)} \circ T_x) f_x\big)$$

$$= \mu_{\theta(x)}\Big(\hat{\mathscr{L}}_x\big((g_{\theta(x)} \circ T_x) f_x\big)\Big) = \mu_{\theta(x)}\big(g_{\theta(x)} \hat{\mathscr{L}}_x(f_x)\big) \tag{3.30}$$

since

$$\hat{\mathscr{L}}_x\big((g_{\theta(x)} \circ T_x) f_x\big) = g_{\theta(x)} \hat{\mathscr{L}}_x(f_x).$$

Substituting in (3.30) $\mathbb{1}_x$ for f_x and using (3.27), we get the lemma. □

Remark 3.10 *In Chap. 4 we provide sufficient measurability conditions for these fiber measures* ν_x *and* μ_x *to be integrable to produce global measures projecting on X to m. The measure* μ *defined by (4.2) is then T-invariant.*

3.5 Levels of Positive Cones of Hölder Functions

For $s \geq 1$, set

$$\Lambda_x^s = \Big\{ g \in \mathscr{C}(\mathscr{J}_x) : g \geq 0, \ \nu_x(g) = 1 \text{ and } g(w_1) \leq e^{sQ_x \varrho^\alpha(w_1, w_2)} g(w_2)$$

$$\text{for all } w_1, w_2 \in \mathscr{J}_x \text{ with } \varrho(w_1, w_2) \leq \xi \Big\}. \tag{3.31}$$

In fact all elements of Λ_x^s belong to $\mathscr{H}^\alpha(\mathscr{J}_x)$. This is proved in the following lemma.

Lemma 3.11 *If* $g \geq 0$ *and if for all* $w_1, w_2 \in \mathscr{J}_x$ *with* $\varrho(w_1, w_2) \leq \xi$ *we have*

$$g(w_1) \leq e^{sQ_x \varrho^\alpha(w_1, w_2)} g(w_2),$$

then

$$v_\alpha(g) \leq sQ_x(\exp(sQ_x\xi^\alpha))\xi^\alpha\|g\|_\infty.$$

Proof. Let $w_1, w_2 \in \mathscr{J}_x$ be such that $\varrho(w_1, w_2) \leq \xi$. Without loss of generality we may assume that $g(w_1) > g(w_2)$. Then $g(w_1) > 0$ and therefore, because of our hypothesis, $g(w_2) > 0$. Hence, we get

$$\frac{|g(w_1) - g(w_2)|}{|g(z_2)|} = \frac{g(w_1)}{g(w_2)} - 1 \leq \exp\left(sQ_x\varrho^\alpha(w_1, w_2)\right) - 1.$$

Then

$$|g(w_1) - g(w_2)| \leq sQ_x(\exp(sQ_x\xi^\alpha))\varrho^\alpha(w_1, w_2)\|g\|_\infty.$$

$\square$

Hence the set Λ_x^s is a level set of the cone defined in (9.13), that is

$$\Lambda_x^s = \mathscr{C}_x^s \cap \{g : v_x(g) = 1\}.$$

In addition, in the following lemma we show that this set is bounded in $\mathscr{H}^\alpha(\mathscr{J}_x)$.

Lemma 3.12 *For a.e. $x \in X$ and every $g \in \Lambda_x^s$, we have $\|g\|_\infty \leq C_{\max}(x)$, where $C_{\max}$ is defined by (3.7).*

Proof. Let $g \in \Lambda_x^s$ and let $z \in \mathscr{J}_x$. Since $g \geq 0$ we get

$$\int_{B(z,\xi)} g \, dv_x \leq \int_{\mathscr{J}_x} g \, dv_x = 1.$$

Therefore there exists $b \in \overline{B}(z, \xi)$ such that

$$g(b) \leq 1/v_x(B(z, \xi)) \leq 1/D_\xi(x),$$

where the latter inequality is due to Lemma 3.6. Hence

$$g(z) \leq e^{sQ_x\varrho^\alpha(b,z)}g(b) \leq \frac{e^{sQ_x}}{D_\xi(x)} \leq C_{\max}(x). \qquad \square$$

A kind of converse to Lemma 3.11 is given by the following.

Lemma 3.13 *If $g \in \mathscr{H}^\alpha(\mathscr{J}_x)$ and $g \geq 0$, then*

$$\frac{g + v_\alpha(g)/Q_x}{v_x(g) + v_\alpha(g)/Q_x} \in \Lambda_x^1.$$

Proof. Consider the function $h = g + v_\alpha(g)/Q_x$. In order to get the inequality from the definition of Λ_x^s, we take $z_1, z_2 \in \mathscr{J}_x$. If $h(z_1) \leq h(z_2)$ then this inequality is trivial. Otherwise $h(z_1) > h(z_2)$, and therefore

$$\frac{h(z_1)}{h(z_2)} - 1 = \frac{|h(z_1) - h(z_2)|}{|h(z_2)|} \leq \frac{v_\alpha(g)\varrho^\alpha(z_1, z_2)}{v_\alpha(g)/Q_x} = Q_x \varrho^\alpha(z_1, z_2). \qquad \square$$

An important property of the sets Λ_x^s is their invariance with respect to the normalized operator $\tilde{\mathscr{L}}_x = \lambda_x^{-1} \mathscr{L}_x$.

Lemma 3.14 *Let $g \in \Lambda_x^s$. Then, for every $n \geq 1$,*

$$\frac{\tilde{\mathscr{L}}_x^n g(w_1)}{\tilde{\mathscr{L}}_x^n g(w_2)} \leq \exp\left(s Q_{x_n} \varrho^\alpha(w_1, w_2)\right), \quad w_1, w_2 \in \mathscr{J}_{\theta^n(x)} \text{ with } \varrho(w_1, w_2) \leq \xi.$$

Consequently, $\tilde{\mathscr{L}}_x^n(\Lambda_x^s) \subset \Lambda_{\theta^n(x)}^s$ for a.e. $x \in X$ and all $n \geq 1$.

Notice that the constant function $\mathbb{1} \in \Lambda_x^s$ for every $s \geq 1$. For this particular function our distortion estimation was already proved in Lemma 3.8.

Proof. [Proof of Lemma 3.14] Let $g \in \Lambda_x^s$, let $\varrho_{\theta^n(x)}(w_1, w_2) \leq \xi$, and let $z_1 \in T_x^{-n}(w_1)$. For $y = (x, z_1)$, we put $z_2 = T_y^{-n}(w_2)$. With this notation, we obtain from Lemma 2.2 and from the definition of Λ_x^s that

$$\frac{\tilde{\mathscr{L}}_x^n g(w_1)}{\tilde{\mathscr{L}}_x^n g(w_2)} \leq \sup_{z_1 \in T_x^{-n}(w_1)} \frac{\exp\left(S_n \varphi_x(z_1)\right) g(z_1)}{\exp\left(S_n \varphi_x(z_2) g(z_2)\right)}$$

$$\leq \exp\left(\varrho^\alpha(w_1, w_2)\left(\sum_{j=0}^{n-1} H_{\theta^j(x)} (\gamma_{\theta^j(x)}^{n-j})^{-\alpha} + s Q_x (\gamma_x^n)^{-\alpha}\right)\right).$$

$$(3.32)$$

Since

$$Q_x(\gamma_x^n)^{-\alpha} + \sum_{j=0}^{n-1} H_{\theta^j(x)}(\gamma_{\theta^j(x)}^{n-j})^{-\alpha} = Q_{\theta^n(x)}, \qquad (3.33)$$

the lemma follows. $\square$

Lemma 3.15 *With $C_{\min}$ the function given by (3.6) we have that*

$$\tilde{\mathscr{L}}_{x-i}^i g \geq C_{\min}(x) \quad \text{for every } i \geq j(x) \text{ and } g \in \Lambda_{x-i}^s.$$

Proof. First, let $i = j(x)$. Since $\int_{\mathscr{J}_{x-i}} g \, dv_{x-i} = 1$ there exists $a \in \mathscr{J}_{x-i}$ such that $g(a) \geq 1$. By definition of $j(x)$, for any point $w \in \mathscr{J}_x$, there exists $z \in T_{x-i}^{-i}(x) \cap B(a, \xi)$. Therefore

$$\tilde{\mathscr{L}}^i_{x-i} g(w) \geq e^{S_i \varphi_{x-i}(z)} g(z) \geq e^{S_i \varphi_{x-i}(z)} e^{-sQ_x} g(a) \geq C_{\min}(x).$$

The case $i > j(x)$ follows from the previous one, since $\tilde{\mathscr{L}}^{i-j(x)}_{x-i} g_{x-i} \in \Lambda_{x-j(x)}$. $\qquad \square$

3.6 Exponential Convergence of Transfer Operators

Lemma 3.16 *Let* $\beta_x = \beta_x(s)$ *(cf. (3.8)). Then for* $x \in X$, $i \geq j(x)$ *and* $g_{x-i} \in \Lambda^s_{x-i}$, *there exists* $h_x \in \Lambda^s_x$ *such that*

$$(\mathscr{L}^i g)_x = \mathscr{L}^i_{x-i} g_{x-i} = \beta_x q_x + (1 - \beta_x) h_x .$$

Proof. By Lemma 3.15, we have $\tilde{\mathscr{L}}^i_{x-i} g_{x-i} \geq C_{\min}(x)$. Then by (3.19) for all $w, z \in \mathscr{J}_x$ with $\varrho_x(z, w) < \xi$,

$$\beta_x \left(\exp\left(sQ_x \varrho^\alpha_x(z, w)\right) q_x(z) - q_x(w) \right) \leq$$

$$\leq \beta_x \left(\exp\left(sQ_x \varrho^\alpha_x(z, w)\right) - \exp\left(-sQ_x \varrho^\alpha_x(z, w)\right) \right) q_x(z)$$

$$\leq \beta_x \left(\exp\left(sQ_x \varrho^\alpha_x(z, w)\right) - \exp\left(-sQ_x \varrho^\alpha_x(z, w)\right) \right) C_\varphi(x)$$

$$\leq \beta_x C_\varphi(x) \left(1 - \exp(-2sQ_x \varrho^\alpha_x(z, w))\right) \exp\left(sQ_x \varrho^\alpha_x(z, w)\right)$$

$$\leq \left(\exp\left(sQ_x \varrho^\alpha_x(z, w)\right) - \exp\left((sQ_x - H_{x-1} \gamma^{-\alpha}_{x-1}) \varrho^\alpha_x(z, w)\right) \right) \tilde{\mathscr{L}}^i_{x-i} g_{x-i}(z)$$

$$\leq \left(\exp\left(sQ_x \varrho^\alpha_x(z, w)\right) - \exp\left((sQ_{x-1} + H_{x-1}) \gamma^{-\alpha}_{x-1} \varrho^\alpha_x(z, w)\right) \right) \tilde{\mathscr{L}}^i_{x-i} g_{x-i}(z).$$

Since by (3.32), for $h \in \Lambda^s_{x-1}$,

$$\tilde{\mathscr{L}}_{x-1} h(z) \leq \exp\left((sQ_{x-1} + H_{x-1}) \gamma^{-\alpha}_{x-1} \varrho^\alpha_x(z, w)\right) \tilde{\mathscr{L}}_{x-1} h(w),$$

$$\tilde{\mathscr{L}}^i_{x-i} g_{x-i}(z) \leq \exp\left((sQ_{x-1} + H_{x-1}) \gamma^{-\alpha}_{x-1} \varrho^\alpha_x(z, w)\right) \tilde{\mathscr{L}}^i_{x-i} g_{x-i}(w).$$

Then we have that

$$\beta_x \left(\exp\left(sQ_x \varrho^\alpha_x(z, w)\right) q_x(z) - q_x(w) \right)$$

$$\leq \exp\left(sQ_x \varrho^\alpha_x(z, w)\right) \tilde{\mathscr{L}}^i_{x-i} g_{x-i}(z) - \tilde{\mathscr{L}}^i_{x-i} g_{x-i}(w)$$

and then

$$\tilde{\mathscr{L}}^i_{x-i} g_{x-i}(w) - \beta_x q_x(w) \leq \exp\left(sQ_x \varrho^\alpha_x(z, w)\right) \left(\tilde{\mathscr{L}}^i_{x-i} g_{x-i}(z) - \beta_x q_x(z) \right).$$

Moreover, $\beta_x q_x \leq C_{\min}(x) \leq \tilde{\mathscr{L}}^i_{x-i} g_{x-i}$. Hence the function

$$h_x := \frac{\tilde{\mathscr{L}}^i_{x-i} g_{x-i} - \beta_x q_x}{1 - \beta_x} \in \Lambda^s_x. \qquad \square$$

We are now ready to establish the first result about exponential convergence.

Proposition 3.17 *Let* $s > 1$. *There exist* $B < 1$ *and a measurable function* $A :$ $X \rightarrow (0, \infty)$ *such that for a.e.* $x \in X$ *for every* $N \geq 1$ *and* $g_{x-N} \in \Lambda^s_{x-N}$ *we have*

$$\|(\tilde{\mathscr{L}}^N g)_x - q_x\|_\infty = \|\mathscr{L}^N_{x-N} g_{x-N} - q_x\|_\infty \leq A(x)B^N.$$

Proof. Fix $x \in X$. Put $g_n := g_{x_n}$, $\beta_n := \beta_{x_n}$, $\Lambda^s_n := \Lambda^s_{x_n}$, and $(\tilde{\mathscr{L}}^n g)_k := (\tilde{\mathscr{L}}^n g)_{x_k}$. Let $(i(n))^\infty_{n=1}$ be a sequence of integers such that $i(n+1) \geq j(x_{-S(n)})$, where $S(n) = \sum^n_{k=1} i(k)$, $n \geq 1$, and where $S(0) = 0$. If $g_{-S(n)} \in \Lambda^s_{-S(n)}$, then Lemma 3.16 yields the existence of a function $h_{n-1} \in \Lambda^s_{-S(n-1)}$ such that

$$\left(\tilde{\mathscr{L}}^{i(n)} g\right)_{-S(n-1)} = \beta_{-S(n-1)} q_{-S(n-1)} + (1 - \beta_{-S(n-1)}) h_{n-1}$$

$$= \left(1 - (1 - \beta_{-S(n-1)})\right) q_{-S(n-1)} + (1 - \beta_{-S(n-1)}) h_{n-1}.$$

Since

$$\left(\tilde{\mathscr{L}}^{i(n)+i(n-1)} g\right)_{-S(n-2)} = \left(\tilde{\mathscr{L}}^{i(n-1)}\left(\tilde{\mathscr{L}}^{i(n)} g\right)\right)_{-S(n-2)}$$

$$= \left(\tilde{\mathscr{L}}^{i(n-1)}\left(\beta_{-S(n-1)} q_{-S(n-1)} + (1 - \beta_{-S(n-1)}) h_{n-1}\right)\right)_{-S(n-2)}$$

$$= \beta_{-S(n-1)} q_{-S(n-2)} + (1 - \beta_{-S(n-1)}) \left(\tilde{\mathscr{L}}^{i(n-1)}(h_{n-1})\right)_{-S(n-2)}$$

it follows again from Lemma 3.16 that there is $h_{n-2} \in \Lambda^s_{-S(n-2)}$ such that

$$\left(\tilde{\mathscr{L}}^{i(n)+i(n-1)} g\right)_{-S(n-2)} = \beta_{-S(n-1)} q_{-S(n-2)}$$

$$+ (1 - \beta_{-S(n-1)})\left(\beta_{-S(n-2)} q_{S(n-2)} + (1 - \beta_{-S(n-2)}) h_{n-2}\right)$$

$$= \left(1 - (1 - \beta_{-S(n-2)})(1 - \beta_{-S(n-1)})\right) q_{S(n-2)}$$

$$+ (1 - \beta_{-S(n-2)})(1 - \beta_{-S(n-1)}) h_{n-1}.$$

It follows now by induction that there exists $h \in \Lambda^s_x$ such that

$$\left(\tilde{\mathscr{L}}^{S(n)} g\right)_x = \left(\tilde{\mathscr{L}}^{i(n)+...+i(1)} g\right)_x = (1 - \Pi^{(n)}_x) q_x + \Pi^{(n)}_x h,$$

where we set $\Pi_x^{(n)} = \prod_{k=0}^{n-1}(1 - \beta_{x-S(k)})$. Since $h \in \Lambda_x^s$, we have $|h| \leq C_{\max}(x)$. Therefore,

$$\left|\left(\tilde{\mathscr{L}}^{S(n)}g\right)_x - \left(1 - \Pi_x^{(n)}\right)q_x\right| \leq C_{\max}(x)\Pi_x^{(n)} \qquad \text{if } g_{-S(n)} \in \Lambda_{-S(n)}^s. \quad (3.34)$$

By measurability of β and j one can find $M > 0$ and $J \geq 1$ such that the set

$$G := \{x : \beta_x \geq M \text{ and } j(x) \leq J\} \quad (3.35)$$

has a positive measure larger than or equal to $3/4$. Now, we will show that for a.e. $x \in X$ there exists a sequence $(n_k)_{k=0}^\infty$ of non-negative integers such that $n_0 = 0$, for $k > 0$, we have that $x_{-Jn_k} \in G$, and

$$\#\{n : 0 \leq n < n_k \text{ and } x_{-Jn} \in G\} = k - 1. \quad (3.36)$$

Indeed, applying Birkhoff's Ergodic Theorem to the mapping θ^{-J} we have that for almost every $x \in X$,

$$\lim_{n\to\infty} \frac{\#\{0 \leq m \leq n - 1 : \theta^{-Jm}(x) \in G\}}{n} = \mathscr{E}(\mathbb{1}_G | \mathscr{I}_J)(x),$$

where $\mathscr{E}(\mathbb{1}_G | \mathscr{I}_J)$ is the conditional expectation of $\mathbb{1}_G$ with the respect to the σ-algebra $\mathscr{I}_J$ of θ^{-J}-invariant sets. Note that if a measurable set A is θ^{-J}-invariant, then set $\cup_{j=0}^{J-1}\theta^j(A)$ is θ^{-1}-invariant. If $m(A) > 0$, then from ergodicity of θ^{-1} we get that $m(\cup_{j=0}^{J-1}\theta^j(A)) = 1$, and then by invariantness of the measure m, we conclude that $m(A) \geq 1/J$. Hence we get that for almost every x the sequence n_k is infinite and

$$\lim_{k\to\infty} \frac{k}{n_k} \geq \frac{3}{4J}. \quad (3.37)$$

Fix $N \geq 0$ and take $l \geq 0$ so that $Jn_l \leq N \leq Jn_{l+1}$. Define a finite sequence $(S(k))_{k=1}^l$ by $S(k) := Jn_k$ for $k < l$ and $S(l) := N$, and observe that by (3.37), we have $N \leq Jn_{l+1} \leq 4J^2l$. Then (3.19) and (3.34) give

$$\|\tilde{\mathscr{L}}_{x-N}^N g_{x-N} - q_x\|_\infty \leq \left\|\tilde{\mathscr{L}}_{x-N}^N g_{x-N} - \left(1 - \Pi_x^{(l)}\right)q_x\right\|_\infty + \Pi_x^{(l)}\|q_x\|_\infty$$

$$\leq (1 - M)^l \left(C_\varphi(x) + C_{\max}(x)\right)$$

$$\leq \left(\sqrt[4J^2]{1 - M}\right)^N \left(C_\varphi(x) + C_{\max}(x)\right).$$

This establishes our proposition with $B = \sqrt[4J^2]{1 - M}$ and

$$A(x) := \max\{2C_{\max}(x)B^{-Jk_x^*}, (C_\varphi(x) + C_{\max}(x))\},$$

where k_x^* is a measurable function such that we have $\frac{k}{n_k} \geq \frac{1}{2J}$ for all $k \geq k_x^*$. $\square$

From now onwards throughout this section, rather than the operator $\mathscr{L}$, we consider the operator $\hat{\mathscr{L}}_x$ defined previously in (3.26).

Lemma 3.18 *Let $s > 1$ and let $g : \mathscr{J} \to \mathbb{R}$ be any function such that $g_x \in \mathscr{H}^\alpha(\mathscr{J}_x)$. Then, with the notation of Proposition 3.17, we have*

$$\left\| \hat{\mathscr{L}}_x^n g_x - \left(\int g_x d\mu_x \right) \mathbb{1} \right\|_\infty \leq C_\varphi(\theta^n(x)) \left(\int |g_x| d\mu_x + 4 \frac{v_\alpha(g_x q_x)}{Q_x} \right) A(\theta^n(x)) B^n.$$

Proof. Fix $s > 1$. First suppose that $g_x \geq 0$. Consider the function

$$h_x = \frac{g_x + v_\alpha(g_x)/Q_x}{\Delta_x} \quad \text{where} \quad \Delta_x := v_x(g_x) + v_\alpha(g_x)/Q_x.$$

It follows from Lemma 3.13 that h_x belongs to the set Λ_x^s and from Proposition 3.17 we have

$$\left\| \tilde{\mathscr{L}}_x^n g_x - \left(\int g_x \, dv_x \right) q_{\theta^n(x)} \right\|_\infty \leq \left\| \Delta_x \tilde{\mathscr{L}}_x^n h_x - \frac{v_\alpha(g_x)}{Q_x} \tilde{\mathscr{L}}_x^n \mathbb{1}_x - \left(\int g_x \, dv_x \right) q_{\theta^n(x)} \right\|_\infty$$

$$= \left\| \Delta_x \tilde{\mathscr{L}}_x^n h_x - \Delta_x q_{\theta^n(x)} + \frac{v_\alpha(g_x)}{Q_x} \left(q_{\theta^n(x)} - \tilde{\mathscr{L}}_x^n \mathbb{1}_x \right) \right\|_\infty$$

$$\leq \left(\Delta_x + \frac{v_\alpha(g_x)}{Q_x} \right) A(\theta^n(x)) B^n.$$

Then applying this inequality for $g_x q_x$ and using (3.19) we get

$$\left\| \hat{\mathscr{L}}_x^n g_x - \left(\int g_x d\mu_x \right) \mathbb{1}_{\theta^n(x)} \right\|_\infty \leq \left\| \frac{1}{q_{\theta^n(x)}} \right\| \cdot \left\| \tilde{\mathscr{L}}_x^n (g_x q_x) - \left(\int g_x q_x \, dv_x \right) q_{\theta^n(x)} \right\|_\infty$$

$$\leq C_\varphi(\theta^n(x)) \left(\int g_x d\mu_x + 2 \frac{v_\alpha(g_x q_x)}{Q_x} \right) A(\theta^n(x)) B^n.$$

So, we have the desired estimate for non-negative g_x. In the general case we can use the standard trick and write $g_x = g_x^+ - g_x^-$, where $g_x^+, g_x^- \geq 0$. Then the lemma follows. $\square$

The estimate obtained in Lemma 3.18 is a little bit inconvenient for it depends on the values of a measurable function, namely $C_\varphi A$, along the positive θ-orbit of $x \in X$. In particular, it is not clear at all from this statement that the item 1 in Theorem 3.2 holds. In order to remedy this flaw, we prove the following proposition.

Proposition 3.19 *For m-a.e. $x \in X$ and every $g_x \in \mathscr{C}(\mathscr{J}_x)$, we have*

$$\left\| \hat{\mathscr{L}}_x^n g_x - \left(\int g_x d\mu_x \right) \mathbb{1}_{\theta^n(x)} \right\|_\infty \xrightarrow[n \to \infty]{} 0.$$

Proof. First of all, we may assume without loss of generality that the function $g_x \in \mathscr{H}^\alpha(\mathscr{J}_x)$ since every continuous function is a limit of a uniformly convergent sequence of Hölder functions. Now, let $\mathscr{A} > 0$ be sufficiently big such that the set

$$X_{\mathscr{A}} = \{x \in X; \ A(x) \leq \mathscr{A}\} \tag{3.38}$$

has positive measure. Notice that, by ergodicity of m, some iterate of a.e. $x \in X$ is in the set $X_{\mathscr{A}}$. Then by Poincaré recurrence theorem and ergodicity of m, for a.e. $x \in X$, there exists a sequence $n_j \to \infty$ such that $\theta^{n_j}(x) \in X_{\mathscr{A}}$, $j \geq 1$. Therefore we get, for such an $x \in X_{\mathscr{A}}$, from Lemma 3.18 that

$$\left\| \hat{\mathscr{L}}_x^{n_j} g_x - \left(\int g_x d\mu_x \right) \mathbb{1}_{\theta^{n_j}(x)} \right\|_\infty \left(\int |g_x| \, d\mu_x + 4 \frac{v_\alpha(g_x q_x)}{Q_x} \right)^{-1} \leq \mathscr{A} B^{n_j} \tag{3.39}$$

for every $j \geq 1$. Finally, to pass from the subsequence (n_j) to the sequence of all natural numbers we employ the monotonicity argument that already appeared in Walters paper [29]. Since $\hat{\mathscr{L}}_x \mathbb{1}_x = \mathbb{1}_{\theta(x)}$, we have for every $w \in \mathscr{J}_{\theta(x)}$ that

$$\inf_{z \in \mathscr{J}_x} g_x(z) \leq \sum_{z \in T_x^{-1}(w)} g_x(z) e^{\hat{\varphi}(z)} \leq \sup_{z \in \mathscr{J}_x} g_x(z).$$

Consequently, the sequence

$$(M_{n,x})_{n=0}^\infty = (\sup_{w \in \mathscr{J}_{\theta^n(x)}} \hat{\mathscr{L}}_x^n g_x(w))_{n=0}^\infty$$

is weakly decreasing. Similarly we have a weakly increasing sequence

$$(m_{n,x})_{n=0}^\infty = (\inf_{w \in \mathscr{J}_{\theta^n(x)}} \hat{\mathscr{L}}_x^n g_x(w))_{n=0}^\infty.$$

The proposition follows since, by (3.39), both sequences converge on the subsequence (n_j). □

3.7 Exponential Decay of Correlations

The following proposition proves item 3 of Theorem 3.2. For a function $f_x \in L^1(\mu_x)$ we denote its L^1-norm with respect to μ_x by

$$\|f_x\|_1 := \int |f_x| d\mu_x.$$

Proposition 3.20 *There exists a θ-invariant set $X' \subset X$ of full m-measure such that, for every $x \in X'$, every $f_{\theta^n(x)} \in L^1(\mu_{\theta^n(x)})$ and every $g_x \in \mathscr{H}^\alpha(\mathscr{J}_x)$,*

$$\left| \mu_x\left((f_{\theta^n(x)} \circ T_x^n) g_x \right) - \mu_{\theta^n(x)}(f_{\theta^n(x)}) \mu_x(g_x) \right| \leq A_*(g_x, \theta^n(x)) B^n \|f_{\theta^n(x)}\|_1,$$

where

$$A_*(g_x, \theta^n(x)) := C_\varphi(\theta^n(x)) \left(\int |g_x| d\mu_x + 4 \frac{v_\alpha(g_x q_x)}{Q_x} \right) A(\theta^n(x)).$$

Proof. Set $h_x = g_x - \int g_x d\mu_x$ and note that by (3.30) and (3.27) we have that

$$\mu_x\left((f_{\theta^n(x)} \circ T_x^n) g_x \right) - \mu_{\theta^n(x)}(f_{\theta^n(x)}) \mu_x(g_x) = \mu_{\theta^n(x)}\left(f_{\theta^n(x)} \hat{\mathscr{L}}_x^n(g_x) \right)$$

$$- \mu_{\theta^n(x)}(f_{\theta^n(x)}) \mu_x(g_x)$$

$$= \mu_{\theta^n(x)}\left(f_{\theta^n(x)} \hat{\mathscr{L}}_x^n(h_x) \right). \quad (3.40)$$

Since Lemma 3.18 yields $\|\hat{\mathscr{L}}_x^n h_x\|_\infty \leq A_*(g_x, \theta^n(x)) B^n$ it follows from (3.40) that

$$\left| \mu_x\left((f_{\theta^n(x)} \circ T_x^n) g_x \right) - \mu_{\theta^n(x)}(f_{\theta^n(x)}) \mu_x(g_x) \right| \leq \int \left| f_{\theta^n(x)} \hat{\mathscr{L}}_x^n(h_x) \right| d\mu_{\theta^n(x)}$$

$$\leq A_*(g_x, \theta^n(x)) B^n \int \left| f_{\theta^n(x)} \right| d\mu_{\theta^n(x)}.$$

$\square$

Using similar arguments like in Proposition 3.19 we obtain the following.

Corollary 3.21 *Let $f_{\theta^n(x)} \in L^1(\mu_{\theta^n(x)})$ and $g_x \in L^1(\mathscr{J}_x)$, where $x \in X'$ and X' is the set given by Lemma 3.20. If $\|f_{\theta^n(x)}\|_1 \neq 0$ for all n, then*

$$\frac{\left| \mu_x\left((f_{\theta^n(x)} \circ T_x^n) g_x \right) - \mu_{\theta^n(x)}(f_{\theta^n(x)}) \mu_x(g_x) \right|}{\|f_{\theta^n(x)}\|_1} \longrightarrow 0 \quad as \ n \to \infty.$$

Remark 3.22 *Note that if $\|f_{\theta^n(x)}\|_1$ grows subexponentially, then*

$$\left| \mu_x\left((f_{\theta^n(x)} \circ T_x^n) g_x \right) - \mu_{\theta^n(x)}(f_{\theta^n(x)}) \mu_x(g_x) \right| \longrightarrow 0 \quad as \ n \to \infty. \quad (3.41)$$

This is for example the case if $x \mapsto \log \|f_x\|_1$ is m-integrable since Birkhoff's Ergodic Theorem implies that $(1/n) \log \|f_{\theta^n(x)}\|_1 \to 0$ for a.e. $x \in X$.

3.8 Uniqueness

Lemma 3.23 *The family of measures $x \mapsto v_x$ is uniquely determined by condition (3.1).*

Proof. Let $\tilde{v}_x$ be a family of probability measures satisfying (3.1). For $x \in X$ choose arbitrarily a sequence of points $w_n \in \mathscr{J}_{\theta^n(x)}$ and define

$$v_{x,n} := \frac{(\mathscr{L}_x^n)^* \delta_{w_n}}{\mathscr{L}_x^n \mathbb{1}(w_n)}.$$

Then, by Proposition 3.19, for a.e. $x \in X$ and all $g_x \in \mathscr{C}(\mathscr{J}_x)$ we have

$$\lim_{n \to \infty} v_{x,n}(g_x) = \lim_{n \to \infty} \frac{\mathscr{L}_x^n g_x(w_n)}{\mathscr{L}_x^n \mathbb{1}(w_n)} = \lim_{n \to \infty} \frac{\hat{\mathscr{L}}_x^n (g_x/q_x)(w_n)}{\hat{\mathscr{L}}_x^n (1/q_x)(w_n)} = \frac{v_x(g_x)}{v_x(\mathbb{1})} = v_x(g_x).$$

$$(3.42)$$

In other words,

$$v_{x,n} \xrightarrow[n \to \infty]{} v_x.$$

$$(3.43)$$

in the weak* topology. Uniqueness of the measures v_x follows. □

Lemma 3.24 *There exists a unique function* $q \in \mathscr{C}^0(\mathscr{J})$ *that satisfies* (3.2).

Proof. Follows from Proposition 3.17. □

3.9 Pressure Function

The pressure function is defined by the formula

$$x \mapsto P_x(\varphi) := \log \lambda_x.$$

If it does not lead to misunderstanding, we will also denote the pressure function by P_x. It is important to note that this function is generally non-constant, even for a.e. $x \in X$. Actually, if the pressure function is a.e. constant, then the random map shares many properties with a deterministic system. This will be explained in detail in Sect. 5. Note that (3.42) and (3.11) imply an alternative definition of $P_x(\varphi)$, namely

$$P_x(\varphi) = \log(v_{\theta(x)}(\mathscr{L}_x \mathbb{1})) = \lim_{n \to \infty} \log \frac{\mathscr{L}_x^{n+1} \mathbb{1}(w_{n+1})}{\mathscr{L}_{\theta(x)}^n \mathbb{1}(w_{n+1})},$$

$$(3.44)$$

where, for every $n \in \mathbb{N}$, w_n is an arbitrary point from $\mathscr{J}_{\theta^n(x)}$.

Lemma 3.25 *For m-a.e.* $x \in X$ *and every sequence* $(w_n)_n \subset \mathscr{J}_x$

$$\lim_{n \to \infty} \frac{1}{n} S_n P_{x-n} - \frac{1}{n} \log \mathscr{L}_{x-n}^n \mathbb{1}_{x-n}(w_n) = 0.$$

Proof. By (3.19) and Proposition 3.17, we have that

$$\frac{1}{C_\varphi(x)} - A(x)B^n \le \frac{\mathscr{L}^n_{x-n} \mathbb{1}_{x-n}(w)}{\lambda^n_{x-n}} \le C_\varphi(x) + A(x)B^n$$

for every $w \in \mathscr{J}_x$ and every $n \in \mathbb{N}$. Therefore

$$\log\left(\frac{1}{C_\varphi(x)} - A(x)\right) \le \log \mathscr{L}^n_{x-n} \mathbb{1}_{x-n}(w) - \log \lambda^n_{x-n} \le \log\left(C_\varphi(x) + A(x)\right).$$

$\square$

Lemma 3.26 *For m-a.e. $x \in X$ and for every sequence $y_n \in \mathscr{J}_{x_n}$, $n \ge 0$,*

$$\underset{n\to\infty}{\text{s-lim}}\left(\frac{1}{n}S_n P_x - \frac{1}{n}\log \mathscr{L}^n_x \mathbb{1}_x(y_n)\right) = 0.$$

Proof. Using Egorov's Theorem and Lemma 3.25 we have that for each $\delta > 0$ there exists a set F_δ such that $m(X \setminus X_\delta) < \delta$ and

$$\frac{1}{n}S_n P_{x-n} - \frac{1}{n}\max_{y\in\mathscr{J}_{xn}}\log \mathscr{L}^n_{x-n}\mathbb{1}_{x-n}(y) \xrightarrow[n\to\infty]{} 0$$

uniformly on F_δ. The lemma follows now from Birkhoff's Ergodic Theorem. $\square$

Lemma 3.27 *If there exist $g \in L^1(m)$ such that $\log\|\mathscr{L}_x\mathbb{1}\|_\infty \le g(x)$, then*

$$\lim_{n\to\infty}\left\|\frac{1}{n}S_n P_x - \frac{1}{n}\log \mathscr{L}^n_x \mathbb{1}_x\right\|_\infty = 0.$$

Proof. Let $F := F_\delta$ be the set from the proof of Lemma 3.26, let $x \in X'_{+F}$ and let (n_j) be the visiting sequence. Let j be such that $n_j < n \le n_{j+1}$. Then

$$\log \mathscr{L}^n_x \mathbb{1}(y) \le \log \|\mathscr{L}^{n_j}_x \mathbb{1}\| + S_{n-n_j} g(\theta^{n_j}(x)) \quad \text{for every } y \in \mathscr{J}_{\theta^n(x)}. \tag{3.45}$$

Now, let $h(x) := \|\varphi_x\|_\infty$. Since by $(3.12) - \log \lambda_x \le \|\varphi_x\|_\infty$,

$$-\log \lambda^n_x = -\log \lambda^{n_j}_x - \log \lambda^{n-n_j}_{x_{n_j}} \le S_{n_j} P_x + S_{n-n_j} h(\theta^{n_j}(x)).$$

Then by (3.45)

$$\frac{1}{n}S_n P_x - \frac{1}{n}\log \mathscr{L}^n_x \mathbb{1}_x(y_{x_n}) \le \frac{1}{n_j}S_{n_j} P_x - \frac{1}{n_j}\log \mathscr{L}^{n_j}_x \mathbb{1}_x(y_{x_{n_j}}) + \frac{1}{n}S_{n-n_j}(g+h)(\theta^{n_j}(x)).$$

On the other hand, for $y \in \mathscr{J}_{\theta^n(x)}$,

$$\log \mathscr{L}^n_x \mathbb{1}(y) \ge \log \mathscr{L}^{n_{j+1}}_x \mathbb{1}(T^{n_{j+1}-n}_{\theta^n(x)}(y)) - S_{n_{j+1}-n} g(\theta^n(x))$$

and by (3.12),

$$\log \lambda_x^n = \log \lambda_x^{n_{j+1}} - \log \lambda_{xn}^{n_{j+1}-n} \leq \log \|\mathscr{L}_x^{n_{j+1}} \mathbb{1}\| + S_{n_{j+1}-n} h(\theta^n(x)).$$

The lemma follows now by Birkhoff's Ergodic Theorem. $\qquad\square$

3.10 Gibbs Property

Lemma 3.28 *Let* $w \in \mathscr{I}_x$, *set* $y = (x, w)$ *and let* $n \geq 0$. *Then*

$$e^{-Q_{\theta^n(x)}}(D_\xi(\theta^n(x))) \leq \frac{\nu_x(T_y^{-n}(B(T^n(y), \xi)))}{\exp(S_n\varphi(y) - S_n P_x(\varphi))} \leq e^{Q_{\theta^n(x)}}.$$

Proof. Fix an arbitrary $z \in \mathscr{I}_x$ and set $y = (x, z)$. Then by Lemma 2.3 and (3.13) we have that

$$\frac{\nu_x(T_y^{-n}(B(T^n(y), \xi)))}{\exp(S_n\varphi(y) - S_n P_x(\varphi))} \leq \frac{(\lambda_x^n)^{-1} \nu_{\theta^n(x)}(B(T^n(y), \xi)) \sup_{z' \in T_y^{-n}(B(T^n(y), \xi))} e^{S_n\varphi(z')}}{(\lambda_x^n)^{-1} e^{S_n\varphi(y)}}$$

$$\leq e^{Q_{S^n(x)}}.$$

On the other hand

$$\frac{\nu_x(T_y^{-n}(B(T^n(y), \xi)))}{\exp(S_n\varphi(y) - S_n P_x(\varphi))} \geq \frac{(\lambda_x^n)^{-1} \nu_{\theta^n(x)}(B(T^n(y), \xi)) \inf_{z' \in T_y^{-n}(B(T^n(y), \xi))} e^{S_n\varphi(z')}}{(\lambda_x^n)^{-1} e^{S_n\varphi(y)}}$$

$$\geq \nu_{\theta^n(x)}(B(T^n(y), \xi)) e^{-Q_{S^n(x)}}.$$

The lemma follows by (3.17). $\qquad\square$

Lemma 3.29 *Let* $T : \mathscr{I} \to \mathscr{I}$ *satisfy the condition of measurability of cardinality of covers and let* $\{\nu_{i,x}\}$, *where* $i = 1, 2$, *be two Gibbs families with pseudo-pressure functions* $x \mapsto P_{i,x}$. *Then, for a.e.* x, *the measures* $\nu_{1,x}$ *and* $\nu_{2,x}$ *are equivalent and*

$$\lim_{k\to\infty} \frac{1}{n_k} S_{n_k} P_{1,x} = \lim_{k\to\infty} \frac{1}{n_k} S_{n_k} P_{2,x} = \lim_{k\to\infty} \frac{1}{n_k} S_{n_k} P_x,$$

where $(n_k) = (n_k(x))$ *is the visiting sequence of an essential set.*

Proof. Let A be compact subset of $\mathscr{I}_x$ and let $\delta > 0$. By regularity of $\nu_{2,x}$ we can find $\varepsilon > 0$ such that

$$\nu_{2,x}(B_x(A, \varepsilon)) \leq \nu_{2,x}(A) + \delta. \tag{3.46}$$

Now, let N_x be a measurable function such that $\xi(\gamma_x^{N_x})^{-1} \leq \varepsilon/2$. Set

$$A_n^j := \{y \in T_x^{-n}(y_{x_n}^j) : A \cap T_y^{-n}(B(y_{x_n}^j, \xi)) \neq \emptyset\}.$$

Let Z be a L, N, D, D-essential set of a_x, N_x, D_1, D_2 and let $(n_k) = (n_k(x))$ be the visiting sequence of Z. Fix $k \in \mathbb{N}$ and put $n = n_k(x)$. Then we have

$$A \subset \bigcup_{j=1}^{a_{x_n}} \bigcup_{y \in A_n^j} T_y^{-n} B(y_{x_n}^j, \xi) \subset B_x(A, \varepsilon).$$

By (3.3) it follows that

$$\nu_{1,x}(A) \leq \sum_{j=1}^{a_{x_n}} \sum_{y \in A_n^j} \nu_{1,x}(T_y^{-n} B(y_{x_n}^j, \xi)) \leq D_1(x) D \sum_{j=1}^{L} \sum_{y \in A_n^j} \exp(S_n\varphi(y) - S_n P_{1,x}(\varphi)).$$

$$(3.47)$$

Then by (3.46) and again by (3.3)

$$\nu_{1,x}(A) \leq D_1(x) D \exp(S_n P_{2,x} - S_n P_{1,x}) \sum_{j=1}^{a_{x_n}} \sum_{y \in A_n^j} \exp(S_n\varphi(y) - S_n P_{2,x}(\varphi))$$

$$\leq D_1(x) D_2(x) D^2 \exp(S_n P_{2,x} - S_n P_{1,x}) \sum_{j=1}^{a_{x_n}} \sum_{y \in A_n^j} \nu_{2,x}(T_y^{-n} B(y_{x_n}^j, \xi))$$

$$\leq D_1(x) D_2(x) D^2 L \exp(S_n P_{2,x} - S_n P_{1,x}) \nu_{2,x}(B(A, \varepsilon))$$

$$\leq D_1(x) D_2(x) D^2 L \exp(S_n P_{2,x} - S_n P_{1,x})(\nu_{2,x}(A) + \delta), \qquad (3.48)$$

since for $y \neq y'$ such that $y, y' \in T_x^{-n}(y_{x_n}^j)$, we have that

$$T_y^{-n} B(y_{x_n}^j, \xi) \cap T_{y'}^{-n} B(y_{x_n}^j, \xi) = \emptyset.$$

Hence the difference $S_{n_k} P_{2,x} - S_{n_k} P_{1,x}$ is bounded from below by some constant, since otherwise taking $A = \mathscr{J}_x$ we would obtain that $\nu_{1,x}(\mathscr{J}_x) = 0$ on a subsequence of (n_k) in (3.48). Similarly, exchanging $\nu_{1,x}$ with $\nu_{2,x}$ we obtain that $S_{n_k} P_{1,x} - S_{n_k} P_{2,x}$ is bounded from above. Then, letting δ go to zero, we have that $\nu_{1,x}$ and $\nu_{2,x}$ are equivalent.

Note that

$$\exp(-S_n P_{1,x}) \mathscr{L}_x^n \mathbb{1}_x(y_n) = \sum_{y \in T_x^{-n}(y_n)} e^{S_n\varphi_x(y) - S_n P_{1,x}}$$

$$\leq D_1(x) D \sum_{y \in T_x^{-n}(y_n)} \nu_{1,x}(T_y^{-n} B(y_n, \xi)) \leq D_1(x) D \nu_{1,x}(\mathscr{J}_x)$$

$$= D_1(x) D.$$

Then

$$\frac{1}{n}\log \mathcal{L}_x 1_x(y_n) - \frac{1}{n}\log(D_1(x)D) \le \frac{1}{n}S_n P_{1,x}.$$

On the other hand, by (3.47), on the same subsequence

$$1 = \nu_x^1(\mathcal{J}_x) \le D_1(x)DL \sum_{y \in T_x^{-n}(y_n)} e^{S_n \varphi_x(y) - S_n P_{1,x}}$$

for some $y_n \in \{y_{x_n}^1, \ldots, y_{x_n}^{a_{x_n}}\}$. Therefore, using Lemma 3.26 and the Sandwich Theorem, we have that, for $x \in X'_Z \cap X'_P$,

$$\lim_{k \to \infty} \frac{1}{n_k} S_{n_k} P_{1,x} = \lim_{k \to \infty} \frac{1}{n_k} S_{n_k} P_x. \qquad \square$$

Remark 3.30 *We cannot expect that $P_{1,x} = P_x(\varphi)$ m-almost surely since, for any measurable function $x \mapsto g_x$, $P_{1,x} := P_x(\varphi) + g_x - g_{\theta(x)}$, is also a pseudo-pressure function (see Lemma 3.28).*

3.11   Some Comments on Uniformly Expanding Random Maps

By $\mathscr{C}_*^\infty(\mathcal{J})$ we denote the space of $\mathcal{B}$-measurable mappings $g : \mathcal{J} \to \mathbb{R}$ with $g_x : \mathcal{J}_x \to \mathbb{R}$ continuous such that $\sup_{x \in X} \|g_x\|_\infty < \infty$. For $H_0 \ge 0$, by $\mathscr{H}_*^\alpha(\mathcal{J}, H_0)$ we denote the space of all functions φ in $\mathscr{H}_m^\alpha(\mathcal{J}) \cap \mathscr{C}_m^\infty(\mathcal{J})$ such that all of H_x are bounded above by H_0. Let

$$\mathscr{H}_*^\alpha(\mathcal{J}) = \bigcup_{H_0 \ge 0} \mathscr{H}_*^\alpha(\mathcal{J}, H_0).$$

For $\varphi \in \mathscr{H}^\alpha(\mathcal{J}, H_0)$ we put

$$Q := H_0 \sum_{j=1}^\infty \gamma^{-\alpha j} = \frac{H_0 \gamma^{-\alpha}}{1 - \gamma^{-\alpha}}.$$

Then Lemma 2.3 takes on the following form.

Lemma 3.31 *For every $\varphi \in \mathscr{H}_*^\alpha(\mathcal{J}, H_0)$,*

$$|S_n \varphi_x(T_y^{-n}(w_1)) - S_n \varphi_x(T_y^{-n}(w_2))| \le Q \varrho^\alpha(w_1, w_2)$$

for all $n \ge 1$, all $x \in X$, every $z \in \mathcal{J}_x$ and every $w_1, w_2 \in B(T^n(z), \xi)$ and where $y = (x, z)$.

In this paper, whenever we deal with uniformly expanding random maps, we always assume that potentials belong to $\mathscr{H}_*^{\alpha}(\mathscr{J})$. Hence all the functions $C_{\varphi}(x)$, $C_{\max}(x)$, $C_{\min}(x)$ and β_x defined, respectively, by (3.5)–(3.8) are uniformly bounded on X. Therefore, there exists $A \in \mathbb{R}$ such that $A(x) \leq A$ for all $x \in X$, where $A(x)$ is the function from Proposition 3.17. In particular, we can prove the following.

Lemma 3.32 *There exists a constant A_{λ} such that, for $x \in X$ and all $y_1, y_2 \in \mathscr{J}_{x_n}$*

$$\left| \frac{\mathscr{L}_x^n \mathbb{1}(y_1)}{\mathscr{L}_{x_1}^{n-1} \mathbb{1}(y_1)} - \lambda_x \right| \leq A_{\lambda} B^n.$$

Proof. It follows from Proposition 3.17 that

$$\left| \tilde{\mathscr{L}}_{x_1}(\tilde{\mathscr{L}} \mathbb{1})(y_1) - \tilde{\mathscr{L}}_{x_1} \mathbb{1}(y_2) \right| \leq 2 A B^{n-1}.$$

Then by Lemma 3.6 and (3.22) we have, for some x-independent constant A_{λ}, that

$$\left| \frac{\mathscr{L}_x^n \mathbb{1}(y_1)}{\mathscr{L}_{x_1}^{n-1} \mathbb{1}(y_2)} - \lambda_x \right| \leq \frac{2 A B^{-1} B^n \lambda_x}{\tilde{\mathscr{L}}_{x_1}^n(\mathbb{1})(y_2)} \leq A_{\lambda} B^n. \qquad \square$$

Chapter 4
Measurability, Pressure and Gibbs Condition

We now study measurability of the objects produced in the previous section. Up to now we do not know, for example, whether the family of measures ν_x represents the disintegration of a global Gibbs state ν with marginal m on the fibered space $\mathscr{J}$. Therefore, we define abstract measurable expanding random maps for which the above measurabilities of λ_x, q_x, ν_x and μ_x can be shown. Then, we can construct a Borel probability invariant ergodic measure on $\mathscr{J}$ for the skew-product transformation T with Gibbs property and study the corresponding expected pressure.

Our settings are related to those of smooth expanding random mappings of one fixed Riemannian manifold from [17] and those of random subshifts of finite type whose fibers are subsets of $\mathbb{N}^{\mathbb{N}}$ from [5]. One possible extension of these works is to consider expanding random transformations on subsets of a fixed Polish space. A general framework for this was, in fact, prepared by Crauel in [10]. In Chap. 4.5 we show how Crauel's random compact subsets of Polish spaces fit into our general framework and, therefore, our settings comprise all these options and go beyond.

The issue of measurability of λ_x, q_x, ν_x and μ_x does not seem to have been treated with care in the literature. As a matter of fact, it was not quite clear to us even for symbol dynamics or random expanding systems of smooth manifolds until, very recently, when Kifer's paper [19] has appeared to take care of these issues.

4.1 Measurable Expanding Random Maps

Let $T : \mathscr{J} \to \mathscr{J}$ be a general expanding random map. Define $\pi_X : \mathscr{J} \to X$ by $\pi_X(x, y) = x$. Let $\mathscr{B} := \mathscr{B}_{\mathscr{J}}$ be a σ-algebra on $\mathscr{J}$ such that

1. π_X and T are measurable,
2. for every $A \in \mathscr{B}$, $\pi_X(A) \in \mathscr{F}$,
3. $\mathscr{B}|_{\mathscr{J}_x}$ is the Borel σ-algebra on $\mathscr{J}_x$.

V. Mayer et al., *Distance Expanding Random Mappings, Thermodynamical Formalism,*
Gibbs Measures and Fractal Geometry, Lecture Notes in Mathematics 2036,
DOI 10.1007/978-3-642-23650-1_4, © Springer-Verlag Berlin Heidelberg 2011

By $L_m^0(\mathscr{J})$ we denote the set of all $\mathscr{B}_\mathscr{J}$-measurable functions and by $\mathscr{C}_m^0(\mathscr{J})$ the set of all $\mathscr{B}_\mathscr{J}$-measurable functions g such that $g_x \in \mathscr{C}(\mathscr{J}_x)$.

Lemma 4.1 *If $g \in \mathscr{C}_m^0(\mathscr{J})$, then $x \mapsto \|g_x\|_\infty$ is measurable.*

Proof. The proof is a consequence of item 2. Indeed, let (G_n) be an increasing approximation of $|g|$ by step functions. So let $G_n = \sum_{k=1}^m a_k \mathbb{1}_{A_k}$, where (a_k) is an increasing sequence of non-negative real numbers, and A_k are $\mathscr{B}_\mathscr{J}$-measurable. Then, define

$$X_m := \pi_X(A_m) \quad \text{and} \quad X_k := \pi_X(A_k) \setminus \cup_{j=k+1}^m \pi_X(A_j),$$

where $k = 1, \ldots, m-1$. Let

$$H_n(x) := \sum_{k=0}^m a_k \mathbb{1}_{X_k}(x) = \sup_{y \in \mathscr{J}_x} G_n(x, y).$$

Then the sequence (H_n) is increasing and converges pointwise to the function $x \mapsto \|g_x\|_\infty$. $\qquad \square$

The space $L_m^1(\mathscr{J})$ is, by definition, the set of all $g \in L_m^0(\mathscr{J})$, such that $\int \|g_x\|_\infty dm(x) < \infty$. We also define

$$\mathscr{C}_m^1(\mathscr{J}) := \mathscr{C}_m^0(\mathscr{J}) \cap L_m^1(\mathscr{J})$$

and

$$\mathscr{H}_m^\alpha(\mathscr{J}) := \mathscr{C}_m^1(\mathscr{J}) \cap \mathscr{H}^\alpha(\mathscr{J}).$$

By $\mathscr{M}^1(\mathscr{J})$ we denote the set of probability measures and by $\mathscr{M}_m^1(\mathscr{J})$ its subset consisting of measures v' such that there exists a system of fiber measures $\{v_x'\}_{x \in X}$ with the property that for every $g \in L_m^1(\mathscr{J})$, the map $x \mapsto \int_{\mathscr{J}_x} g_x\, dv_x'$ is measurable and

$$\int_\mathscr{J} g dv' = \int_X \int_{\mathscr{J}_x} g_x dv_x' dm(x).$$

Then

$$m = v' \circ \pi_X^{-1} \tag{4.1}$$

and the family $(v_x')_{x \in X}$ is the canonical system of conditional measures of v' with respect to the measurable partition $\{\mathscr{J}_x\}_{x \in X}$ of $\mathscr{J}$. It is also instructive to notice that in the case when $\mathscr{J}$ is a Lebesgue space then (4.1) implies that $v' \in \mathscr{M}_m^1(\mathscr{J})$.

The measure $\mu' \in \mathscr{M}^1(\mathscr{J})$ is called *T-invariant* if $\mu' \circ T^{-1} = \mu'$. If $\mu' \in \mathscr{M}_m^1(\mathscr{J})$, then, in terms of the fiber measures, clearly *T*-invariance equivalently means that the family $\{\mu_x'\}_{x \in X}$ is *T*-invariant; see Chap. 3.1 for the definition of *T*-invariance of a family of measures.

Fix $\varphi \in \mathscr{H}_m^1(\mathscr{J})$. Then the general expanding random map $T : \mathscr{J} \to \mathscr{J}$ is called *a measurable expanding random map* if the following conditions are satisfied.

Measurability of the Transfer Operator. The transfer operator is *measurable*, i.e. $\mathscr{L}g \in \mathscr{C}_m^0(\mathscr{J})$ for every $g \in \mathscr{C}_m^0(\mathscr{J})$.

Integrability of the Logarithm of the Transfer Operator. The function $X \ni x \mapsto \log\|\mathscr{L}_x \mathbb{1}_x\|_\infty$ belongs to $L^1(m)$.

We shall now provide a simple, easy to verify, sufficient condition for integrability of the logarithm of the transfer operator.

Lemma 4.2 *If* $\log(\deg(T_x)) \in L^1(m)$*, then* $x \mapsto \log\|\mathscr{L}_x \mathbb{1}_x\|_\infty$ *belongs to* $L^1(m)$.

Proof. Recall that

$$e^{-\|\varphi_x\|_\infty} \leq \sum_{T_x(z)=w} e^{\varphi_x(z)} \leq \deg(T_x)e^{\|\varphi_x\|_\infty}.$$

Hence $-\|\varphi_x\|_\infty \leq \log\|\mathscr{L}_x \mathbb{1}_x\|_\infty \leq \log(\deg(T_x)) + \|\varphi_x\|_\infty.$ $\qquad\square$

4.2 Measurability

Now, we assume that $T : \mathscr{J} \to \mathscr{J}$ is a measurable expanding random map. In particular, the operator $\mathscr{L}$ is measurable. Armed with these assumptions, we come back to the families of Gibbs states $\{v_x\}_{x\in X}$ and $\{\mu_x\}_{x\in X}$ whose pointwise construction was given in Theorem 3.1. Since we have already established good convergence properties, especially the exponential decay of correlations, it will follow rather easily that these families form in fact conditional measures of some measures v and μ from $\mathscr{M}_m^1(\mathscr{J})$. As an immediate consequence of item 3 of Theorem 3.1, we get that the probability measure μ is invariant under the action of the map $T : \mathscr{J} \to \mathscr{J}$. All of this is shown in the following lemmas.

Lemma 4.3 *For every* $g \in L_m^1(\mathscr{J})$*, the map* $x \mapsto v_x(g_x)$ *is measurable.*

Proof. It follows from (3.42) that

$$\lim_{n\to\infty} \frac{\|\mathscr{L}_x^n g_x\|_\infty}{\|\mathscr{L}_x^n \mathbb{1}\|_\infty} = v_x(g_x).$$

Then measurability of $x \mapsto v_x(g_x)$ is a direct consequence of measurability of the transfer operator. $\qquad\square$

This lemma enables us to introduce the probability measure v on $\mathscr{J}$ given by the formula:

$$v(g) = \int_X \int_{\mathscr{J}_x} g_x \, dv_x \, dm(x).$$

This measure, therefore, belongs to $\mathscr{M}_m^1(\mathscr{J})$.

Lemma 4.4 *The map $X \ni x \mapsto \lambda_x \in \mathbb{R}$ is measurable and the function $q : \mathscr{J} \ni$*
$(x, y) \mapsto q_x(y)$ belongs to $L^0_m(\mathscr{J})$.

Proof. Since $v \in \mathcal{M}^1_m(\mathscr{J})$, measurability of λ's follows from the formula (3.11)
and measurability of the transfer operator. Then measurability of λ's and of the
transfer operator together with $\lim_{n\to\infty} \tilde{\mathscr{L}}^n_{x-n} \mathbb{1} = q_x$ (see Proposition 3.17) imply
measurability of q.   □

From this lemma and Lemma 4.3 it follows that we can define a measure μ by
the formula:

$$\mu(g) = \int_X \int_{\mathscr{J}_x} q_x g_x dv_x dm(x). \tag{4.2}$$

4.3 The Expected Pressure

The pressure function of a measurable expanding random map has the following
important property.

Lemma 4.5 *The pressure function $X \ni x \mapsto P_x(\varphi)$ is integrable.*

Proof. It follows from the definition of the transfer operator, that

$$-\|\varphi_x\|_\infty \le \log v_{\theta(x)}(\mathscr{L}_x \mathbb{1}) \le \log \|\mathscr{L}_x \mathbb{1}\|_\infty. \tag{4.3}$$

Then, by (3.11) and integrability of the logarithm of the transfer operator, the function $P_x(\varphi)$ is bounded above and below by integrable functions, hence integrable.
                                                                                                        □

Therefore, *the expected pressure* of φ given by

$$\mathscr{E}P(\varphi) = \int_X P_x(\varphi) dm(x)$$

is well defined.

The equality (3.42) yields alternative formulas for the expected pressure. In order
to establish them, observe that by Birkhoff's Ergodic Theorem

$$\mathscr{E}P(\varphi) = \lim_{n\to\infty} \frac{1}{n} \log \lambda^n_x \quad \text{for a.e. } x \in X. \tag{4.4}$$

In addition, by (3.11), $\lambda^n_x = \lambda^n_x v_x(\mathbb{1}) = v_{\theta^n(x)}(\mathscr{L}^n_x(\mathbb{1}))$. Thus, it follows that

$$\frac{1}{n} \log \lambda^n_x = \lim_{k\to\infty} \frac{1}{n} \log \frac{\mathscr{L}^{k+n}_x \mathbb{1}_x(w_{k+n})}{\mathscr{L}^k_{\theta^n(x)} \mathbb{1}_{\theta^n(x)}(w_{k+n})}.$$

However, by Lemma 3.27 we can get even more interesting formula.

Lemma 4.6 *For every $\varphi \in \mathscr{H}_m^\alpha(\mathscr{J})$ and for almost every $x \in X$*

$$\mathscr{E}P(\varphi) = \lim_{n\to\infty} \frac{1}{n} \log \mathscr{L}_x^n \mathbb{1}(w_n),$$

where the points $w_n \in \mathscr{J}_{\theta^n(x)}$ are arbitrarily chosen.

4.4 Ergodicity of μ

Proposition 4.7 *The measure μ is ergodic.*

Proof. Let B be a measurable set such that $T^{-1}(B) = B$ and, for $x \in X$, denote by B_x the set $\{y \in \mathscr{J}_x : (x, y) \in B\}$. Then we have that $T_x^{-1}(B_{\theta(x)}) = B_x$. Now let

$$X_0 := \{x \in X : \mu_x(B_x) > 0\}.$$

This is clearly a θ-invariant subset of X. We will show that, if $m(X_0) > 0$, then $\mu_x(B_x) = 1$ for a.e. $x \in X_0$. Since θ is ergodic with respect to m, this implies ergodicity of T with respect to μ.

Define a function f by $f_x := \mathbb{1}_{B_x}$. Clearly $f_x \in L^1(\mu_x)$ and $f_{\theta^n(x)} \circ T_x^n = f_x$ m-a.e. Let $x \in X' \cap X_0$, where X' is given by Proposition 3.20. Let g_x be a function from $L^1(\mathscr{J}_x)$ with $\int g_x d\mu_x = 0$. Then using (3.41) we obtain that

$$\lim_{n\to\infty} \mu_x\big((f_{\theta^n(x)} \circ T_x^n) g_x\big) \to 0.$$

Consequently

$$\int_{B_x} g_x \, d\mu_x = 0.$$

Since this holds for every mean zero function $g_x \in L^1(\mathscr{J}_x)$, we have that $\mu_x(B_x) = 1$ for every $x \in X' \cap X_0$. This finishes the proof of ergodicity of T with respect to the measure μ. $\qquad\square$

A direct consequence of Lemma 3.29 and ergodicity of T is the following.

Proposition 4.8 *The measure $\mu \in \mathscr{M}_m^1(\mathscr{J})$ is a unique T-invariant measure satisfying (3.3).*

4.5 Random Compact Subsets of Polish Spaces

Suppose that $(X, \mathscr{F}, m)$ is a complete measure space. Suppose also that (Y, ϱ) is a Polish space which is normalized so that $\mathrm{diam}(Y) = 1$. Let $\mathscr{B}_Y$ be the σ-algebra of Borel subsets of Y and let $\mathscr{K}_Y$ be the space of all compact subsets of Y topologized by the Hausdorff metric. Assume that a measurable mapping $X \ni x \mapsto \mathscr{J}_x \in \mathscr{K}_Y$ is given.

Following Crauel [10, Chap. 2], we say that a map $X \ni x \mapsto Y_x \subset Y$ is *measurable* if for every $y \in Y$, the map $x \mapsto d(y, Y_x)$ is measurable, where

$$d(y, Y_x) := \inf\{d(y, y_x) : y_x \in Y_x\}.$$

This map is also called *a random set*. If every Y_x is closed (res. compact), it is called *a closed (res. compact) random set*. With this terminology $X \ni x \mapsto \mathscr{J}_x \subset Y$ is a compact random set (see [10, Remark 2.16, p. 16]).

Closed random sets have the following important properties (cf. [10, Proposition 2.4 and Theorem 2.6]).

Theorem 4.9 *Suppose that $X \ni x \mapsto Y_x$ is a closed random set such that $Y_x \neq \emptyset$.*

(a) *For all open sets $V \subset Y$, the set $\{x \in X : Y_x \cap V \neq \emptyset\}$ is measurable.*
(b) *The set $\mathscr{J} := \mathrm{graph}(x \mapsto Y_x) := \{(x, y_x) : x \in X \text{ and } y_x \in Y_x\}$ is a measurable subset of $X \times Y$, i.e. $\mathscr{J}$ is a subset of $\mathscr{F} \otimes \mathscr{B}_Y$, the product σ-algebra of $\mathscr{F}$ and $\mathscr{B}_Y$.*
(c) *For every n, there exists a measurable function $X \ni x \mapsto y_{x,n} \in Y_x$ such that*

$$Y_x = \mathrm{cl}\{y_{x,n} : n \in \mathbb{N}\}.$$

In particular, there exists a measurable map $X \ni x \mapsto y_x \in Y_x$.

Note that item (b) implies that $\mathscr{J}$ is a measurable subset of $X \times Y$. Let $\mathscr{B}_{\mathscr{J}} := \mathscr{F} \otimes \mathscr{B}_Y|_{\mathscr{J}}$. Then by Theorem 2.12 from [10] we get that for all $A \in \mathscr{B}_{\mathscr{J}}$, $\pi_X(A) \in \mathscr{F}$.

Now, let $X \ni x \mapsto Y_x$ be a compact random set and let $r > 0$ be a real number. Then every set Y_x can be covered by some finite number $a_x = a_x(r) \in \mathbb{N}$ of open balls with radii equal to r. Moreover, by Lebesgue's Covering Lemma, there exits $R_x = R_x(r) > 0$ such that every ball $B(y_x, R_x)$ with $y_x \in Y_x$ is contained in a ball from this cover. As we prove below, we can actually choose a_x and R_x in a measurable way. Hence for the compact random set $x \mapsto \mathscr{J}_x$ the measurability of cardinality of covers (see Chap. 3.1, just before Theorem 3.3) holds automatically.

In the proof of Lemma 4.11 we will use the following Proposition 2.1 from [10, p. 15].

Proposition 4.10 *For compact random set $x \mapsto Y_x$ and for every ε, there exists a (non-random) compact set $Y_\varepsilon \subset Y$ such that*

$$m(\{x \in X : Y_x \subset Y_\varepsilon\}) \geq 1 - \varepsilon.$$

Lemma 4.11 *There exists a measurable set $X'_a \subset X$ of full measure m such that, for every $r > 0$ and every positive integer k, there exists a measurable function $X'_a \ni x \mapsto y_{x,k} \in Y_x$ and there exist measurable functions $X'_a \ni x \mapsto a_x \in \mathbb{N}$ and $X'_a \ni x \mapsto R_x \in \mathbb{R}_+$ such that for every $x \in X'_a$,*

$$\bigcup_{k=1}^{a_x} B_x(y_{x,k}, r) \supset Y_x,$$

and for every $y_x \in Y_x$, *there exists* $k = 1, \ldots, a_x$ *for which* $B_x(y_x, R_x) \subset B_x(y_{x,k}, r)$.

Proof. For $n \in \mathbb{N}$ let $Y_{1/n} \subset Y$ be a compact set given by Proposition 4.10. Then the set $X_n := \{x \in X : Y_x \subset Y_{1/n}\}$ is measurable and has the measure $m(X_n)$ greater or equal to $1 - 1/n$. Define

$$X'_a := \bigcup_{n \in \mathbb{N}} X_n.$$

Then $m(X'_a) = 1$.

Let $\{y_n : n \in \mathbb{N}_+\}$ be a dense subset of Y. Since $Y_{1/n}$ is compact, there exists a positive integer $a(n)$ such that

$$\bigcup_{k=1}^{a(n)} B(y_k, r/2) \supset Y_{1/n}. \tag{4.5}$$

Define a function $X'_a \ni x \mapsto a_x$, by $a_x = a(n)$ where $n := \min\{k : x \in X_k\}$. The measurability of X_n gives us the required measurability of $x \mapsto a_x$.

Let $\{y_k : k \in \mathbb{N}\}$ be a countable dense set of Y and $m \in \mathbb{N}$. For every $k \in \mathbb{N}$ define a function $x \mapsto G_{x,k}$ by

$$G_{x,k} = \begin{cases} \overline{B}(y_k, r/2) & \text{if } Y_x \cap B(y_k, r/2) \neq \emptyset \\ Y_x & \text{otherwise.} \end{cases}$$

Since, by Theorem 4.9(a), the set $\{x \in X : Y_x \cap B(y_k, r/2) \neq \emptyset\}$ is measurable, it follows that $X \ni x \mapsto G_{x,k}$ is a closed random set. Hence, by Theorem 4.9(c), there exists a measurable selection $X \ni x \mapsto y_{x,k} \in G_{x,k}$. Note that, if $y_{x,k} \in \overline{B}(y_k, r/2)$, then $B(y_k, r/2) \subset B(y_{x,k}, r)$. Therefore, by (4.5),

$$\bigcup_{k=1}^{U_x} B(y_{x,k}, r) \supset Y_{1/n} \supset Y_x \quad \text{for all } x \in X_n.$$

Finally, for $x \in X_n$, let $R_x > 0$ be a real number such that, for $y \in Y_{1/n}$, there exists $k = 1, \ldots, U(n)$ for which $B(y, R_x) \subset B(y_k, r/2) \subset B(y_{x,k}, r)$. Then $X'_U \ni x \mapsto R_x \in \mathbb{R}_+$ is also measurable. $\square$

Chapter 5
Fractal Structure of Conformal Expanding Random Repellers

We now deal with *conformal expanding random maps*. We prove an appropriate version of Bowen's Formula, which asserts that the Hausdorff dimension of almost every fiber $\mathscr{J}_x$, denoted throughout the paper by HD, is equal to a unique zero of the function $t \mapsto \mathscr{E}P(t)$. We also show that typically Hausdorff and packing measures on fibers respectively vanish and are infinite. A simple example of such a phenomenon is a Random Cantor Set described.

Later in this paper the reader will find more refined and general examples of Random Conformal Systems notably Classical Random Expanding Systems, Brück and Büger Polynomial Systems and DG-Systems.

In the following we suppose that all the fibers $\mathscr{J}_x$ are in an ambient space Y which is a smooth Riemannian manifold. We will deal with $C^{1+\alpha}$-conformal mappings f_x and denote then $|f'_x(z)|$ the norm of the derivative of f_x which, by conformality, is nothing else than the similarity factor of $f'_x(z)$. Finally, let $\|f'_x\|_\infty$ be the supremum of $|f'_x(z)|$ over $z \in \mathscr{J}_x$. Since we deal with expanding systems we have

$$|f'_x| \geq \gamma_x \quad \text{for a.e. } x \in X. \tag{5.1}$$

Definition 5.1 *Let* $f : (x,z) \mapsto (\theta(x), f_x(z))$ *be a measurable expanding random map having fibers* $J_x \subset Y$ *and such that the mappings* $f_x : \mathscr{J}_x \to \mathscr{J}_{\theta(x)}$ *can be extended to a neighborhood of* $\mathscr{J}_x$ *in* Y *to conformal* $C^{1+\alpha}$ *mappings. If in addition* $\log \|f'_x\|_\infty \in L^1(m)$ *then we call* f conformal expanding random map.

A conformal random map $f : \mathscr{J} \to \mathscr{J}$ *which is uniformly expanding is called* conformal uniformly expanding.

5.1 Bowen's Formula

For every $t \in \mathbb{R}$ we consider the potential $\varphi_t(x,z) = -t \log |f'_x(z)|$. The associated topological pressure $P(\varphi_t)$ will be denoted $P(t)$. Let

V. Mayer et al., *Distance Expanding Random Mappings, Thermodynamical Formalism, Gibbs Measures and Fractal Geometry*, Lecture Notes in Mathematics 2036, DOI 10.1007/978-3-642-23650-1_5, © Springer-Verlag Berlin Heidelberg 2011

$$\mathscr{E}P(t) = \int_X P_x(t) dm(x)$$

be its expected value with respect to the measure m. In view of (5.1), it follows from Lemma 9.6 that the function $t \mapsto \mathscr{E}P(t)$ has a unique zero. Denote it by h. The result of this subsection is the following version of Bowen's formula, identifying the Hausdorff dimension of almost all fibers with the parameter h.

Theorem 5.2 (Bowen's Formula) *Let f be a conformal expanding random map. The parameter h, i.e. the zero of the function $t \mapsto \mathscr{E}P(t)$, is m-a.e. equal to the Hausdorff dimension* $\mathrm{HD}(\mathscr{J}_x)$ *of the fiber* $\mathscr{J}_x$.

Bowen's formula has been obtained previously in various settings first by Kifer [18] and then by Crauel and Flandoli [11], Bogenschütz and Ochs [6], and Rugh [26].

Proof. Let $(v_{x,h})_{x \in X}$ be the measures produced in Theorem 3.1 for the potential φ_h. Fix $x \in X$ and $z \in \mathscr{J}_x$ and set again $y = (x, z)$. For every $r \in (0, \xi]$ let $k = k(z, r)$ be the largest number $n \geq 0$ such that

$$B(z, r) \subset f_y^{-n}(B(f_x^n(z), \xi)). \tag{5.2}$$

By the expanding property this inclusion holds for all $0 \leq n \leq k$ and $\lim_{r \to 0} k(z, r) = +\infty$. Fix such an n. By Lemma 3.28,

$$v_{x,h}(B(z, r)) \leq v_{x,h}(f_y^{-n}(B(f_x^n(z), \xi))) \leq \exp(hQ_{\theta^n(x)})|(f_x^n)'(z)|^{-h} \exp(-P_x^n(h)). \tag{5.3}$$

On the other hand, $B(z, r) \not\subset f_y^{-(s+1)}(B(f_x^{s+1}(z), \xi))$ for every $s \geq k$. But, since by Lemma 2.3,

$$B(z, \exp(-Q_{\theta^{s+1}(x)}\xi^\alpha)|(f_x^{s+1})'(z)|^{-1}\xi) \subset f_y^{-(s+1)}(B(f_x^{s+1}(z), \xi)),$$

we get

$$\exp(-Q_{\theta^{s+1}(x)}\xi^\alpha)|(f_x^{s+1})'(z)|^{-1}\xi \leq r \tag{5.4}$$

and $|(f_x^s)'(z)|^{-1} \leq \xi^{-1} \exp(Q_{\theta^{s+1}(x)}\xi^\alpha)r$. Inserting this to (5.3) we obtain,

$$v_{x,h}(B(z, r)) \leq \xi^{-h} \exp(hQ_{\theta^n(x)}) \exp(hQ_{\theta^{s+1}(x)}\xi^\alpha)r^h$$
$$\times \exp(-P_x^n(h))|(f_{\theta^n(x)}^{s+1-n})'(f_x^n(z))|^h \tag{5.5}$$

or, equivalently,

$$\frac{\log v_{x,h}(B(z, r))}{\log r} \geq h + \frac{hQ_{\theta^n(x)}}{\log r} + \frac{hQ_{\theta^{s+1}(x)}\xi^\alpha}{\log r} + \frac{-h \log\left(\left|(f_{\theta^n(x)}^{s+1-n})'(f_x^n(z))\right|\right)}{\log r}$$
$$+ \frac{-h \log \xi}{\log r} + \frac{-P_x^n(h)}{\log r}. \tag{5.6}$$

Our goal is to show that

$$\liminf_{r \to 0} \frac{\log v_{x,h}(B(z,r))}{\log r} \geq h \quad \text{for a.e. } x \in X \text{ and all } z \in \mathscr{J}_x.$$

Since the function $x \mapsto Q_x$ is measurable and almost everywhere finite, there exists $M > 0$ such that $m(A) > 0$, where $A = \{x \in X : Q_x \leq M\}$. Fix $n = n_k \geq 0$ to be the largest integer less than or equal to k such that $\theta^n(x) \in A$ and $s = s_k$ to be the least integer greater than or equal to k such that $\theta^{s+1}(x) \in A$. It follows from Birkhoff's Ergodic Theorem that $\lim_{k \to \infty} s_k / n_k = 1$. Of course if for $k \geq 1$ we take any $r_k > 0$ such that $k(z, r_k) = k$, then $\lim_{k \to \infty} r_k = 0$.

Now, note that by (5.2), the formula

$$f_y^{-n}(B(f_x^n(z), \xi)) \subset B(z, \exp(Q_{\theta^n(x)}\xi^\alpha)|(f_x^n)'(z)|^{-1}\xi)$$

yields $r \leq \exp(Q_{\theta^n(x)}\xi^\alpha)|(f_x^n)'(z)|^{-1}\xi$. Equivalently,

$$-\log r \geq \log|(f_x^n)'(z)| - \xi^\alpha Q_{\theta^n(x)} - \log \xi.$$

Since $\log|(f_x^n)'(z)| \geq \log \gamma_x^n$ and since the function $x \mapsto \log \gamma_x$ is integrable and

$$\chi = \min\{1, \int \log \gamma \, dm\} > 0$$

we get from Birkhoff's Ergodic Theorem that for a.e. $x \in X$ and all $r > 0$ small enough (so k and n_k and s_k large enough too)

$$-\log r \geq \frac{\chi}{2}n \geq \frac{\chi}{3}s. \tag{5.7}$$

Remember that $\theta^n(x) \in A$ and $\theta^{s+1}(x) \in A$. We thus obtain from (5.6) that

$$\liminf_{r \to 0} \frac{\log v_{x,h}(B(z,r))}{\log r} \geq h - 3h \limsup_{k \to \infty} \frac{1}{s} \log \left(\left| (f_{\theta^n(x)}^{s+1-n})'(f_x^n(z)) \right| \right) - 2\frac{1}{n}P_x^n(h) \tag{5.8}$$

for a.e. $x \in X$ and all $z \in \mathscr{J}_x$. But as $\int P_x(h)dm(x) = 0$, we have by Birkhoff's Ergodic Theorem that

$$\lim_{n \to \infty} \frac{1}{n}P_x^n(h) = 0. \tag{5.9}$$

Also, since the measure μ_h is f-invariant, it follows from Birkhoff's Ergodic Theorem that there exists a measurable set $X_0 \subset X$ such that for every $x \in X_0$ there exists at least one (in fact of full measure $\mu_{x,h}$) $z_x \in \mathscr{J}_x$ such that

$$\lim_{j \to \infty} \frac{1}{j} \log \left| (f_x^j)'(z_x) \right| = \hat{\chi} := \int_{\mathscr{J}} \log|f_x'(z)|d\mu_h(x,z) \in (0, +\infty).$$

Hence, remembering that $\theta^n(x)$ and $\theta^{s+1}(x)$ belong to A, we get

$$
\begin{aligned}
\limsup_{k\to\infty} \frac{1}{s}\log\left(\left|(f^{s+1-n}_{\theta^n(x)})'(f^n_x(z))\right|\right) &= \limsup_{k\to\infty}\frac{1}{s}\left(\log\left|(f^{s+1}_x)'(z)\right| - \log\left|(f^n_x)'(z)\right|\right)\\
&= \limsup_{k\to\infty}\frac{1}{s}\left(\log\left|(f^{s+1}_x)'(z_x)\right| - \log\left|(f^n_x)'(z_x)\right|\right)\\
&\le \limsup_{k\to\infty}\frac{1}{s}\log\left|(f^{s+1}_x)'(z_x)\right|\\
&\quad - \liminf_{k\to\infty}\frac{1}{s}\log\left|(f^n_x)'(z_x)\right| = \hat{\chi} - \hat{\chi} = 0.
\end{aligned}
$$

Inserting this and (5.9) to (5.8) we get that

$$
\liminf_{r\to 0}\frac{\log \nu_{x,h}(B(z,r))}{\log r} \ge h. \tag{5.10}
$$

Keep $x \in X, z \in \mathscr{J}_x$ and $r \in (0,\xi]$. Now, let $l = l(z,r)$ be the least integer ≥ 0 such that

$$
f^{-l}_y(B(f^l_x(z),\xi)) \subset B(z,r). \tag{5.11}
$$

Then, by Lemma 3.28,

$$
\begin{aligned}
\nu_{x,h}(B(z,r)) &\ge \nu_{x,h}(f^{-l}_y(B(f^l_x(z),\xi)))\\
&\ge D_1(\theta^l(x))\exp(-Q_{\theta^l(x)})|(f^l_x)'(z)|^{-l}\exp(-P^l_x(h)).
\end{aligned} \tag{5.12}
$$

On the other hand, $f^{-(l-1)}_y(B(f^{l-1}_x(z),\xi)) \not\subset B(z,r)$. But, since

$$
f^{-(l-1)}_y(B(f^{l-1}_x(z),\xi)) \subset B(y,\exp(Q_{\theta^{l-1}}(x)\xi^\alpha)|(f^{l-1}_x)'(z)|^{-1}\xi),
$$

we get

$$
r \le \xi\exp(Q_{\theta^{l-1}(x)}\xi^\alpha)|(f^{l-1}_x)'(y)|^{-1}. \tag{5.13}
$$

Thus $|(f^{l-1}_x)'(z)|^{-1} \ge \xi^{-1}\exp(-Q_{\theta^{l-1}(x)}\xi^\alpha)r$. Inserting this to (5.12) we obtain,

$$
\nu_{x,h}(B(z,r)) \ge \xi^{-h}D_1(\theta^l(x))e^{-Q_{\theta^l(x)}}|(f_{\theta^{l-1}(x)})'(f^{l-1}_x(z))|^{-h} \tag{5.14}
$$

$$
\cdot\exp(-hQ_{\theta^{l-1}(x)}\xi^\alpha)r^h\exp(-P^l_x(h)). \tag{5.15}
$$

Now, given any integer $j \ge 1$ large enough, take $R_j > 0$ to be the least radius $r > 0$ such that

$$
f^{-j}_y(B(f^j_x(z),\xi)) \subset B(z,r).
$$

Then $l(y,R_j) = j$. Since the function Q is measurable and almost everywhere finite, and θ is a measure-preserving transformation, there exist a set $\Gamma \subset X$ with positive measure m and a constant $E > 0$ such that $Q_x \le E$, $D_1(x) \le E$ and $Q_{\theta^{-1}(x)} \le E$ for all $x \in \Gamma$. It follows from Birkhoff's Ergodic Theorem and ergodicity of the map $\theta : X \to X$ that there exists a measurable set $X_1 \subset X$ with $m(X_1) = 1$ such that for every $x \in X_1$ there exists an unbounded increasing sequence $(j_i)_{i=1}^\infty$ such that $\theta^{j_i}(x) \in \Gamma$ for all $i \ge 1$. Formula (5.13) then yields

$$-\log R_{j_i} \ge -E\xi^\alpha + \log \xi + \log |(f_x^{j_i-1}(z)| \ge -E\xi^\alpha + \log \xi + \log \gamma_x^{j_i-1} \ge \frac{\chi}{2} j_i,$$

where the last inequality was written because of the same argument as (5.7) was, intersecting also X_1 with an appropriate measurable set of measure 1. Now we get from (5.14) that

$$\frac{\log \nu_{x,h}(B(z, R_{j_i}))}{\log R_{j_i}} \le h + \frac{2\log E}{\chi j_i} - \frac{2E}{\chi j_i} - \frac{2h}{\chi} \frac{1}{j_i} \log \|(f_{\theta^{j_i-1}(x)})'\|_\infty - \frac{2h\xi^\alpha E}{\chi j_i}$$

$$-\frac{2h\log \xi}{\chi j_i} - \frac{2}{\chi} \frac{1}{j_i} P_x^{j_i}(h).$$

Noting that $\int_X P_x(t)dm(x) = 0$ and applying Birkhoff's Ergodic Theorem, we see that the last term in the above estimate converges to zero. Also $\frac{1}{j_i} \log \|(f_{\theta^{j_i-1}(x)})'\|_\infty$ converges to zero because of Birkhoff's Ergodic Theorem and integrability of the function $x \mapsto \log \|f_x'\|_\infty$. Since all the other terms obviously converge to zero, we thus get for a.e. $x \in X$ and all $z \in \mathscr{J}_x$, that

$$\liminf_{r \to 0} \frac{\log \nu_{x,h}(B(z, r))}{\log r} \le \liminf_{i \to \infty} \frac{\log \nu_{x,h}(B(z, R_{j_i}))}{\log R_{j_i}} \le h.$$

Combining this with (5.10), we obtain that

$$\liminf_{r \to 0} \frac{\log \nu_{x,h}(B(z, r))}{\log r} = h$$

for a.e. $x \in X$ and all $z \in \mathscr{J}_x$. This gives that $\mathrm{HD}(\mathscr{J}_x) = h$ for a.e. $x \in X$. We are done. $\qquad\square$

5.2 Quasi-Deterministic and Essential Systems

We now investigate the fractal structure of the Julia sets and we will see that the random systems naturally split into two classes depending on the asymptotic behavior of Birkhoff's sums of the topological pressure $P_x^n(h)$.

Definition 5.3 *Let f be a conformal uniformly expanding random map. It is called essentially random if for m-a.e. $x \in X$,*

$$\limsup_{n\to\infty} P_x^n(h) = +\infty \quad \text{and} \quad \liminf_{n\to\infty} P_x^n(h) = -\infty, \tag{5.16}$$

where h is the Bowen's parameter coming from Theorem 5.2. The map f is called quasi-deterministic if for m-a.e. $x \in X$ there exists $L_x > 0$ such that

$$-L_x \le P_x^n(h) \le L_x \quad \text{for m-almost all } x \in X \text{ and all } n \ge 0. \tag{5.17}$$

Remark 5.4 *Because of ergodicity of the transformation* $\theta : X \to X$, *for a uniformly conformal random map to be essential it suffices to know that the condition (5.16) is satisfied for a set of points* $x \in X$ *with a positive measure m.*

Remark 5.5 *If the number*

$$\sigma^2(P(h)) = \lim_{n\to\infty} \frac{1}{n} \int \left(S_n(P(h))\right)^2 dm > 0$$

and if the Law of Iterated Logarithm holds, i.e. if

$$-\sqrt{2\sigma^2(P(h))} = \liminf_{n\to\infty} \frac{P_x^n(h)}{\sqrt{n \log\log n}} \le \limsup_{n\to\infty} \frac{P_x^n(h)}{\sqrt{n \log\log n}} = \sqrt{2\sigma^2(P(h))}$$

m-a.e., then our conformal random map is essential. It is essential even if only the Central Limit Theorem holds, i.e. if

$$m\left(\left\{x \in X : \frac{P_x^n(h)}{\sqrt{n}} < r\right\}\right) \to \frac{1}{\sigma\sqrt{2\pi}} \int_{-\infty}^{r} e^{-s^2/2\sigma^2(P(h))} \, ds.$$

Remark 5.6 *If there exists a bounded everywhere defined measurable function u :* $X \to \mathbb{R}$ *such that* $P_x(h) = u(x) - u \circ \theta(x)$ *(i.e. if* $P(h)$ *is a coboundary) for all* $x \in X$, *then our system is quasi-deterministic.*

For every $\alpha > 0$ let $\mathscr{H}^\alpha$ refer to the α-dimensional Hausdorff measure and let $\mathscr{P}^\alpha$ refer to the α-dimensional packing measure. Recall that a Borel probability measure μ defined on a metric space M is geometric with an exponent α if and only if there exist $A \ge 1$ and $R > 0$ such that

$$A^{-1}r^\alpha \le \mu(B(z,r)) \le Ar^\alpha$$

for all $z \in M$ and all $0 \le r \le R$. The most significant basic properties of geometric measures are the following:
(GM1) The measures μ, $\mathscr{H}^\alpha$, and $\mathscr{P}^\alpha$ are all mutually equivalent with Radon–Nikodym derivatives separated away from zero and infinity.
(GM2) $0 < \mathscr{H}^\alpha(M), \mathscr{P}^\alpha(M) < +\infty$.
(GM3) $\mathrm{HD}(M) = h$.
    The main result of this section is the following.

Theorem 5.7 *Suppose* $f : \mathscr{J} \to \mathscr{J}$ *is a conformal uniformly expanding random map.*

(a) *If the system* $f : \mathscr{J} \to \mathscr{J}$ *is essential, then* $\mathscr{H}^h(\mathscr{J}_x) = 0$ *and* $\mathscr{P}^h(\mathscr{J}_x) = +\infty$ *for m-a.e.* $x \in X$.

(b) *If, on the other hand, the system* $f : \mathscr{J} \to \mathscr{J}$ *is quasi-deterministic, then for every* $x \in X$ v_x^h *is a geometric measure with exponent h and therefore* (GM1)–(GM3) *hold.*

Proof. Part (a). Remember that by its very definition $\mathscr{E}P(h) = \int P_x(h) dm(x) = 0$. By Definition 5.3 there exists a measurable set X_1 with $m(X_1) = 1$ such that for every $x \in X_1$ there exists an increasing unbounded sequence $(n_j)_{j=1}^{\infty}$ (depending on x) of positive integers such that

$$\lim_{j \to \infty} P_x^{n_j}(h) = -\infty. \tag{5.18}$$

Since we are in the uniformly expanding case, the formula (5.12) from the proof of Theorem 5.2 (Bowen's Formula) takes on the following simplified form:

$$v_x(B(z,r)) \geq D^{-1} r^h \exp\left(-P_x^{l(z,r)}(h)\right) \tag{5.19}$$

with some $D \geq 1$ and all $z \in \mathscr{J}_x$. Since the map is uniformly expanding, for all $j \geq 1$ large enough, there exists $r_j > 0$ such that $l(z,r_j) = n_j$. So disregarding finitely many terms, we may assume without loss of generality, that this is true for all $j \geq 1$. Clearly $\lim_{j \to \infty} r_j = 0$. It thus follows from (5.19) that

$$v_{x,h}(B(z,r_j)) \geq D^{-1} r_j^h \exp\left(-P_x^{n_j}(h)\right)$$

for all $x \in X_1$, all $z \in \mathscr{J}_x$ and all $j \geq 1$. Therefore, by (5.18),

$$\limsup_{r \to 0} \frac{v_{x,h}(B(z,r))}{r^h} \geq \limsup_{j \to \infty} \frac{v_{x,h}(B(z,r_j))}{r_j^h} \geq D^{-1} \limsup_{j \to \infty} \exp\left(-P_x^{n_j}(h)\right) = +\infty,$$

which implies that $\mathscr{H}^h(\mathscr{J}_x) = 0$.

The proof for packing measures is similar. By Definition 5.3 there exists a measurable set X_2 with $m(X_2) = 1$ such that for every $x \in X_2$ there exists an increasing unbounded sequence $(s_j)_{j=1}^{\infty}$ (depending on x) of positive integers such that

$$\lim_{j \to \infty} P_x^{s_j}(h) = +\infty. \tag{5.20}$$

Since we are in the expanding case, formula (5.5) from the proof of Theorem 5.2 (Bowen's Formula), applied with $s = k(z,r)$, takes on the following simplified form:

$$v_x(B(z,r)) \leq D r^h \exp\left(-P_x^{k(z,r)}(h)\right) \tag{5.21}$$

with $D \geq 1$ sufficiently large, all $x \in X_2$ and all $z \in \mathcal{J}_x$. By our uniform assumptions, for all $j \geq 1$ large enough, there exists $R_j > 0$ such that $k(z, R_j) = s_j$. Clearly $\lim_{j \to \infty} R_j = 0$. It thus follows from (5.21) that

$$\nu_{x,h}(B(z, r_j)) \leq DR_j^h \exp\left(-P_x^{s_j}(h)\right)$$

for all $x \in X_2$, all $z \in \mathcal{J}_x$ and all $j \geq 1$. Therefore, using (5.20), we get

$$\liminf_{r \to 0} \frac{\nu_{x,h}(B(z, r))}{r^h} \leq \liminf_{j \to \infty} \frac{\nu_{x,h}(B(z, R_j))}{R_j^h} \leq D \liminf_{j \to \infty} \exp\left(-P_x^{s_j}(h)\right) = 0.$$

Thus $\mathscr{P}^h(\mathcal{J}_x) = +\infty$. We are done with part (a).

Suppose now that the map $f : \mathcal{J} \to \mathcal{J}$ is quasi-deterministic. It then follows from Definition 5.3 and (5.19) along with (5.21), that for every $x \in X$ and for every $r > 0$ small enough independently of $x \in X$, we have

$$(L_x D)^{-1} r^h \leq \nu_{x,h}(B(y, r)) \leq L_x D r^h, \quad x \in X, z \in \mathcal{J}_x.$$

This means that each $\nu_{x,h}$, $x \in X$, is a geometric measure with exponent h and the theorem follows.   □

As a straightforward consequence of this theorem we get a corollary transparently stating that essential conformal random systems are entirely new objects, drastically different from deterministic self-conformal sets.

Corollary 5.8 *Suppose that conformal random map $f : \mathcal{J} \to \mathcal{J}$ is essential. Then for m-a.e. $x \in X$ the following hold.*

(1) *The fiber $\mathcal{J}_x$ is not bi-Lipschitz equivalent to any deterministic nor quasi-deterministic self-conformal set.*
(2) *$\mathcal{J}_x$ is not a geometric circle nor even a piecewise smooth curve.*
(3) *If $\mathcal{J}_x$ has a non-degenerate connected component (for example if $\mathcal{J}_x$ is connected), then $h = \mathrm{HD}(\mathcal{J}_x) > 1$.*
(4) *Let d be the dimension of the ambient Riemannian space Y. Then $\mathrm{HD}(\mathcal{J}_x) < d$.*

Proof. Item (1) follows immediately from Theorem 5.7(a) and (b3). Item (3) from Theorem 5.7(a) and the observation that $\mathscr{H}^1(W) > 0$ whenever W is connected. The proof of (4) is similar. Since (3) obviously implies (2), we are done.   □

5.3 Random Cantor Set

Here is a first example of an essentially random system. Define

$$f_0(x) = 3x \,(\mathrm{mod}\, 1) \quad \text{for } x \in [0, 1/3] \cup [2/3, 1]$$

and

$$f_1(x) = 4x \pmod 1 \text{ for } x \in [0, 1/4] \cup [3/4, 1].$$

Let $X = \{0, 1\}^{\mathbb{Z}}$, θ be the shift transformation and m be the standard Bernoulli measure. For $x = (\ldots, x_{-1}, x_0, x_1, \ldots) \in X$ define $f_x = f_{x_0}$, $f_x^n = f_{\theta^{n-1}(x)} \circ f_{\theta^{n-2}(x)} \circ \ldots \circ f_x$ and

$$\mathscr{J}_x = \bigcap_{n=0}^{\infty} (f_x^n)^{-1}([0, 1]).$$

The skew product map defined on $\bigcup_{x \in X} J_x$ by the formula

$$f(x, y) = (\theta(x), f_x(y))$$

generates a conformal random expanding system. We shall show that this system is essential. To simplify the next calculation, we define recurrently:

$$\xi_x(1) = \begin{cases} 3 & \text{if } x_0 = 0 \\ 4 & \text{if } x_0 = 1 \end{cases}, \ \xi_x(n) = \xi_{\theta^{n-1}(x)}(1)\xi_x(n-1).$$

Consider the potential φ^t defined by the formula $\varphi_x^t = -t \log \xi_x(1)$. Then

$$S_n \varphi_x^t = -t \log \xi_x(n).$$

Let C_n be a cylinder of the order n that is C_n is a subset of $\mathscr{J}_x$ of diameter $(\xi_x(n))^{-1}$ such that $f_x^n|_{C_n}$ is one-to-one and onto $\mathscr{J}_{\theta^n(x)}$. We can project the measure m on $\mathscr{J}_x$ and we call this measure μ_x. In other words, μ_x is such a measure that all cylinders of level n have the measure $1/2^n$. Then by Law of Large Numbers for m-almost every x

$$\lim_{n \to \infty} \frac{\log \mu_x(C_n)}{\log \operatorname{diam}(C_n)} = \frac{\log 2}{(1/n) \log \xi_x(n)} = \frac{\log 4}{\log 12} =: h.$$

Therefore the Hausdorff dimension of $\mathscr{J}_x$ is for m-almost every x constant and equal to h. Next note that

$$\frac{\mu_x(C_n)}{\operatorname{diam}(C_n)^h} = \exp(-S_n P_x), \tag{5.22}$$

where

$$P_x := \log 2 - h \log \xi_x(1).$$

This will give us the value of the Hausdorff and packing measure. So let $Z_0, Z_1, \ldots$ be independent random variables, each having the same distribution such that the probability of $Z_n = \log 2 - h \log 3$ is equal to the probability of $Z_n = \log 2 - h \log 4$

and is equal to $1/2$. The expected value of Z_n, $\mathscr{E}P$, is zero and its standard deviation $\sigma > 0$. Then the Law of the Iterated Logarithm tells us that the following equalities:

$$\liminf_{n \to \infty} \frac{Z_1 + \ldots + Z_n}{\sqrt{n \log \log n}} = -\sqrt{2}\sigma \quad \text{and} \quad \limsup_{n \to \infty} \frac{Z_1 + \ldots + Z_n}{\sqrt{n \log \log n}} = \sqrt{2}\sigma \quad (5.23)$$

hold with probability one. Then, by (5.22),

$$\limsup_{n \to \infty} \frac{\mu_x(C_n)}{\text{diam}(C_n)^h} = \infty \quad \text{and} \quad \liminf_{n \to \infty} \frac{\mu_x(C_n)}{\text{diam}(C_n)^h} = 0$$

for m-almost every x. In particular, the Hausdorff measure of almost every fiber $\mathscr{J}_x$ vanishes and the packing measure is infinite. Note also that the Hausdorff dimension of fibers is not constant as clearly $\text{HD}(\mathscr{J}_{0\infty}) = \log 2/\log 3$, whereas $\text{HD}(\mathscr{J}_{1\infty}) = \log 2/\log 4 = 1/2$.

Chapter 6
Multifractal Analysis

The second direction of our study of fractal properties of conformal random expanding maps is to investigate the multifractal spectrum of Gibbs measures on fibers. We show that the multifractal formalism is valid. It seems that it is impossible to do it with a method inspired by the proof of Bowen's formula since one gets full measure sets for each real α and not one full measure set X_{ma} such that for all $x \in X_{ma}$, the multifractal spectrum of the Gibbs measure on the fiber over x is given by the Legendre transform of a temperature function which is independent of $x \in X_{ma}$. In order to overcome this problem we work out a different proof in which we minimize the use Birkhoff's Ergodic Theorem and instead we base the proof on the definition of Gibbs measures and the behavior of the Perron–Frobenius operator. In this point we were partially motivated by the approach presented in Falconer's book [15].

Another issue we would like to bring up here is real analyticity of the multifractal spectrum which we establish for uniformly expanding systems. The proof is based on real-analiticity results for the expected pressure which are treated separately in Chap. 9 since this part involves different methods.

6.1 Concave Legendre Transform

Let $\varphi \in H_m(\mathcal{J})$ be such that $\mathcal{E}P(\varphi) = 0$. Fix $q \in \mathbb{R}$. We will not use the function q_x and therefore this will not cause any confusion. Define auxiliary potentials

$$\varphi_{q,x,t}(y) := q(\varphi_x(y) - P_x(\varphi)) - t \log |f_x'(y)|.$$

By Lemma 9.5, the function $(q,t) \mapsto \mathcal{E}P(q,t) := \mathcal{E}P(\varphi_{q,t})$ is convex. Moreover, since $\log |f_x'(y)| \geq \log \gamma_x > 0$, it follows from Lemma 9.6 that for every $q \in \mathbb{R}$ there exists a unique $T(q) \in \mathbb{R}$ such that

V. Mayer et al., *Distance Expanding Random Mappings, Thermodynamical Formalism, Gibbs Measures and Fractal Geometry*, Lecture Notes in Mathematics 2036, DOI 10.1007/978-3-642-23650-1_6, © Springer-Verlag Berlin Heidelberg 2011

$$\mathscr{E}P(\varphi_{q,T(q)}) = 0.$$

The function $q \mapsto T(q)$ defined implicitly by this formula is referred to as *the temperature function*. Put

$$\varphi_q := \varphi_{q,T(q)}.$$

By D_T we denote the set of differentiability points of the temperature function T. By convexity of $\mathscr{E}P$, for $\lambda \in (0,1)$,

$$\mathscr{E}P(\lambda q_1 + (1-\lambda)q_2, \lambda T(q_1) + (1-\lambda)T(q_2))$$
$$\leq \lambda \mathscr{E}P(q_1, T(q_1)) + (1-\lambda)\mathscr{E}P(q_2, T(q_2)) = 0.$$

Since $t \mapsto \mathscr{E}P(\lambda q_1 + (1-\lambda)q_2, t)$ is decreasing,

$$T(\lambda q_1 + (1-\lambda)q_2) \leq \lambda T(q_1) + (1-\lambda)T(q_2).$$

Hence the function $q \mapsto T(q)$ is convex and continuous. Furthermore, it follows from its convexity that the function T is differentiable everywhere but a countable set, where it is left and right differentiable. Define

$$L(T)(\alpha) := \inf_{-\infty < q < \infty} \big(\alpha q + T(q)\big),$$

where

$$\alpha \in \mathrm{Dom}(L) = \Big[\lim_{q \to -\infty} -T'(q^-), \lim_{q \to \infty} -T'(q^+)\Big].$$

We call L *the concave Legendre transform*. This transform is related to the (classical) Legendre transform **L** by the formula $L(T)(\alpha) = -\mathbf{L}(T)(-\alpha)$. The transform L sends convex functions to concave ones and, if $q \in D_T$, then

$$L(T)(-T'(q)) = -T'(q)q + T(q).$$

Lemma 6.1 *Let $q \in D_T$. Then for every $\varepsilon > 0$ there exists $\delta_\varepsilon > 0$, such that, for all $\delta \in (0, \delta_\varepsilon)$, we have*

$$\mathscr{E}P((1+\delta)q, T(q) + (qT'(q) + \varepsilon)\delta) < 0$$

and

$$\mathscr{E}P((1-\delta)q, T(q) + (-qT'(q) + \varepsilon)\delta) < 0.$$

Proof. Since the temperature function T is differentiable at the point q, we may write

$$T(q + \delta q) = T(q) + T'(q)\delta q + o(\delta)$$

for all $\delta > 0$ sufficiently small, say $\delta \in (0, \delta_\varepsilon^{(1)})$. So,

$$T(q) + (qT'(q) + \varepsilon)\delta - T((1 + \delta)q) = \varepsilon\delta + o(\delta) > 0.$$

Then, in virtue of Lemma 9.6, we get that

$$\mathscr{E}P((1 + \delta)q, T(q) + (qT'(q) + \varepsilon)\delta) < \mathscr{E}P((1 + \delta)q, T((1 + \delta)q)) = 0,$$

meaning that the first assertion of our lemma is proved. The second one is proved similarly producing a positive number $\delta_\varepsilon^{(2)}$. Setting then $\delta_\varepsilon = \min\{\delta_\varepsilon^{(1)}, \delta_\varepsilon^{(2)}\}$ completes the proof. $\qquad\square$

6.2 Multifractal Spectrum

Let μ be the invariant Gibbs measure for φ and let ν be the φ-conformal measure. For every $\alpha \in \mathbb{R}$ define

$$K_x(\alpha) := \left\{ y \in \mathscr{J}_x : d_{\mu_x}(y) := \lim_{r \to 0} \frac{\log \mu_x(B(y, r))}{\log r} = \alpha \right\}$$

and

$$K'_x := \left\{ y \in \mathscr{J}_x : \text{the limit } \lim_{r \to 0} \frac{\log \mu_x(B(y, r))}{\log r} \text{ does not exist} \right\}.$$

This gives us the multifractal decomposition

$$\mathscr{J}_x := \biguplus_{\alpha \geq 0} K_x(\alpha) \uplus K'_x.$$

The multifractal spectrum is the family of functions $\{g_{\mu_x}\}_{x \in X}$ given by the formulas:

$$g_{\mu_x}(\alpha) := \text{HD}(K_x(\alpha)).$$

The function $d_{\mu_x}(y)$ is called the local dimension of the measure μ_x at the point y. Since for m almost every $x \in X$ the measures μ_x and ν_x are equivalent with Radon–Nikodym derivatives uniformly separated from 0 and infinity (though the bounds may and usually do depend on x), we conclude that we get the same set $K_x(\alpha)$ if in its definition the measure μ_x is replaced by ν_x. Our goal now is to get a "smooth" formula for g_{μ_x}.

Let μ_q and ν_q be the measures for the potential φ_q given by Theorem 3.1. The main technical result of this section is this.

Proposition 6.2 *For every $q \in D_T$ there exists a measurable set $X_{ma} \subset X$ with $m(X_{ma}) = 1$ and such that, for every $x \in X_{ma}$, and all $q \in D_T$, we have*

$$g_{\mu_x}(-T'(q)) = -qT'(q) + T(q)$$

Proof. Firstly, by Lemma 9.4, for every $0 < R \leq \xi$ there exists a measurable function $D_R : X \rightarrow (0, +\infty)$ such that for all $q \in \mathbb{R}$, all $x \in X$, all $y \in \mathscr{J}_x$, and all integers $n \geq 0$, we have

$$D_R^{-q^*}(\theta^n(x)) \leq \frac{v_{q,x}(f_y^{-n}(B(f^n(y), R)))}{\exp(q(S_n\varphi(y) - P_x^n(\varphi)))|(f_x^n)'(y)|^{-T(q)}} \leq D_R^{q^*}(\theta^n(x)),$$

$$(6.1)$$

where $q^* := (q, T(q))^*$ as defined in (9.1). In what follows we keep the notation from the proof of Theorem 5.2. The formulas (5.2) and (5.11) then give for every $j \geq l$ and every $0 \leq i \leq k$, that

$$D_\xi^{-q^*}(\theta^j(x)))^{-1} \exp(q(S_j\varphi(y) - P_x^j(\varphi)))|(f_x^j)'(y)|^{-T(q)}$$

$$\leq v_{q,x}(B(y,r)) \qquad\qquad\qquad (6.2)$$

$$\leq D_\xi^{q^*}(\theta^i(x)) \exp(q(S_i\varphi(y) - P_x^i(\varphi)))|(f_x^i)'(y)|^{-T(q)}.$$

By Q_x we denote the measurable function given by Lemma 2.3 for the function $-\log|f'|$. Let X_* be an essential set for the functions $X \ni x \mapsto R_x$, $X \ni x \mapsto a(x)$, $x \mapsto Q_x$, and $X \ni x \mapsto D_\xi(x)$ with constants $\hat{R}$, $\hat{a}$, $\hat{Q}$, and $\hat{D}_\xi$. Let $(n_j)_1^\infty$ be the positively visiting sequence for X_* at x. Let $X_\mathscr{E}'$ be the set given by Lemma 9.5 for potentials $\phi_{q,t}$, $q, t \in \mathbb{R}^2$. Let

$$X_+' := X_\mathscr{E}' \cap X_{+X_*}'.$$

Let us first prove the upper bound on $g_{\mu_x}(-T'(q))$. Fix $x \in X_+'$. Fix $\varepsilon_1 > 0$. For every $j \geq 1$ let $\{w_k(x_{n_j}) : 1 \leq k \leq a(x_{n_j})\}$ be a ξ spanning set of $\mathscr{J}_{x_{n_j}}$. As $\mathscr{E}P(\phi_q) = 0$, it follows from Lemma 9.6 that $\gamma := \frac{1}{2}\mathscr{E}P(\phi_{q,T(q)+\varepsilon_1}) < 0$. So, in virtue of Lemma 9.5, there exists $C \geq 1$ such that

$$\mathscr{L}_{\phi_{q,T(q)+\varepsilon_1},x}\mathbb{1}(w_k(x_{n_j})) \leq Ce^{-\gamma n_j} \qquad\qquad (6.3)$$

for all $j \leq 1$ and all $k = 1, 2, \ldots, a(\theta^{n_j}(x)) \leq \hat{a}$. Now, fix an arbitrary $\varepsilon_2 \in \mathbb{R}$ such that $q\varepsilon_2 \geq 0$. For every integer $l \geq 1$ let

$$K_x(\varepsilon_2, l) = \left\{ y \in K_x(-T'(q)) : -T'(q) - \frac{1}{2}|\varepsilon_2| \leq \frac{\log v_x(B(y,r))}{\log r} \leq -T'(q) + \frac{1}{2}|\varepsilon_2| \right.$$

$$\left. \text{for all } 0 < r \leq 1/l \right\}.$$

Note that

$$K_x(-T'(q)) = \bigcup_{l=1}^{\infty} K_x(\varepsilon_2, l). \qquad (6.4)$$

Let

$$\Gamma_{n_j}(x) = \left\{ z \in \bigcup_{k=1}^{a(x_{n_j})} f_x^{-n_j}(w_k(x_{n_j})) : K_x(\varepsilon_2, l) \cap f_z^{-n_j}(B(f^{n_j}(z), \xi/2)) \neq \emptyset \right\}.$$

Then

$$K_x(\varepsilon_2, l) \subset \bigcup_{z \in \Gamma_{n_j}(x)} f_z^{-n_j}(B(f^{n_j}(z), \xi/2)). \qquad (6.5)$$

For every $z \in \Gamma_{n_j}(x)$, say $z \in f_x^{-n_j}(w_k(x_{n_j}))$, choose

$$\hat{z} \in K_x(\varepsilon_2, l) \cap f_z^{-n_j}(B(w_k(x_{n_j}), \xi/2)).$$

Then $B(w_k(x_{n_j}), \xi/2) \subset B(f^{n_j}(z), \xi)$, and therefore

$$f_z^{-n_j}(B(w_k(x_{n_j}), \xi/2)) \subset f_{\hat{z}}^{-n_j}(B(f^{n_j}(\hat{z}), \xi)).$$

It follows from this and (6.5) that

$$K_x(\varepsilon_2, l) \subset \bigcup_{z \in \Gamma_{n_j}(x)} f_{\hat{z}}^{-n_j}(B(f^{n_j}(\hat{z}), \xi)). \qquad (6.6)$$

Put

$$r_j^{(1)}(\hat{z}) = \hat{Q}^{-1}|(f_x^{n_j})'(\hat{z})|^{-1} \quad \text{and} \quad r_j^{(2)}(\hat{z}) = \hat{Q}|(f_x^{n_j})'(\hat{z})|^{-1}$$

We then have

$$B(\hat{z}, r_j^{(1)}(\hat{z})) \subset f_{\hat{z}}^{-n_j}(B(f^{n_j}(\hat{z}), \xi)) \subset B(\hat{z}, r_j^{(2)}(\hat{z})).$$

Therefore, assuming $j \geq 1$ to be sufficiently large so that the radii $r_j^{(1)}(\hat{z})$ and $r_j^{(1)}(\hat{z})$ are sufficiently small, particularly $\leq 1/l$, we get

$$\frac{\log \nu_x(f_{\hat{z}}^{-n_j}(B(f^{n_j}(\hat{z}), \xi)))}{-\log |(f_x^{n_j})'(\hat{z})|} \leq \frac{\log \nu_x(B(\hat{z}), \hat{Q}^{-1}|(f_x^{n_j})'(\hat{z})|^{-1})}{-\log |(f_x^{n_j})'(\hat{z})|}$$

$$\leq \frac{\log \nu_x(B(\hat{z}), r_j^{(1)}(\hat{z}))}{\log(r_j^{(1)}(\hat{z})) + \log \hat{Q}} \leq -T'(q) + |\varepsilon_2|$$

and

$$\frac{\log v_x\big(f_{\hat{z}}^{-n_j}\,(B(f^{n_j}(\hat{z}),\xi))\big)}{-\log|(f_x^{n_j})'(\hat{z})|} \geq \frac{\log v_x\big(B(\hat{z}),\hat{Q}|(f_x^{n_j})'(\hat{z})|^{-1}\big)}{-\log|(f_x^{n_j})'(\hat{z})|}$$

$$\geq \frac{\log v_x\big(B(\hat{z}),r_j^{(2)}(\hat{z})\big)}{\log(r_j^{(2)}(\hat{z})) - \log\hat{Q}} \geq -T'(q) - |\varepsilon_2|.$$

Hence,

$$|q|\big(\log v_x\big(f_{\hat{z}}^{-n_j}\,(B(f^{n_j}(\hat{z}),\xi))\big) - (T'(q)+|\varepsilon_2|)\log|(f_x^{n_j})'(\hat{z})|\big) \leq 0$$

and

$$|q|\big(\log v_x\big(f_{\hat{z}}^{-n_j}\,(B(f^{n_j}(\hat{z}),\xi))\big) - (T'(q)-|\varepsilon_2|)\log|(f_x^{n_j})'(\hat{z})|\big) \geq 0.$$

So, in either case (as $\varepsilon_2 q > 0$),

$$-q\big(\log v_x\big(f_{\hat{z}}^{-n_j}\,(B(f^{n_j}(\hat{z}),\xi))\big) - (T'(q)-|\varepsilon_2|)\log|(f_x^{n_j})'(\hat{z})|\big) \leq 0$$

or equivalently,

$$v_x^{-q}\big(f_{\hat{z}}^{-n_j}\,(B(f^{n_j}(\hat{z}),\xi))\big)|(f^{n_j})'(\hat{z})|^{qT'(q)-\varepsilon_2 q} \leq 1. \qquad (6.7)$$

Put $t = -qT'(q) + T(q) + \varepsilon_1 + \varepsilon_2 q$. Using (6.7) and (6.3) we can then estimate as follows:

$$\sum_{z\in\Gamma_{n_j}(x)} \mathrm{diam}^{-qT'(q)+T(q)+\varepsilon_1+\varepsilon_2 q}\big(f_{\hat{z}}^{-n_j}\,(B(f^{n_j}(\hat{z}),\xi))\big)$$

$$= \sum_{z\in\Gamma_{n_j}(x)} \mathrm{diam}^{T(q)+\varepsilon_1}\big(f_{\hat{z}}^{-n_j}\,(B(f^{n_j}(\hat{z}),\xi))\big)\mathrm{diam}^{-qT'(q)+\varepsilon_2 q}\big(f_{\hat{z}}^{-n_j}\,(B(f^{n_j}(\hat{z}),\xi))\big)$$

$$\leq \sum_{z\in\Gamma_{n_j}(x)} (\hat{Q}\xi^{-1})^t|(f^{n_j})'(z)|^{-(T(q)+\varepsilon_1)}(\hat{Q}\xi)^{-t}|(f^{n_j})'(\hat{z})|^{qT'(q)-\varepsilon_2 q}$$

$$= (\hat{Q}\xi^{-1})^{2t} \sum_{z\in\Gamma_{n_j}(x)} \exp\big(q(S_{n_j}\varphi(z) - P_x^{n_j}(\varphi)) - (T(q)+\varepsilon_1)\log|(f_x^{n_j})'(z)|\big)$$

$$\cdot \exp\big(q(P_x^{n_j}(\varphi) - S_{n_j}\varphi(z))|(f^{n_j})'(\hat{z})|^{qT'(q)-\varepsilon_2 q}$$

$$\leq (\hat{Q}\xi^{-1})^{2t}e^{q\hat{Q}_\phi} \sum_{z\in\Gamma_{n_j}(x)} (\hat{Q}\xi^{-1})^{2t} \sum_{z\in\Gamma_{n_j}(x)} \exp\big(q(S_{n_j}\varphi(z) - P_x^{n_j}(\varphi))$$

$$-(T(q)+\varepsilon_1)\log|(f_x^{n_j})'(z)|\big)\exp\big(q(P_x^{n_j}(\varphi) - S_{n_j}\varphi(\hat{z}))\big)|(f^{n_j})'(\hat{z})|^{qT'(q)-\varepsilon_2 q}$$

$$\leq (\hat{Q}\xi^{-1})^{2t} e^{q\hat{Q}_\phi} \sum_{z\in\Gamma_{n_j}(x)} (\hat{Q}\xi^{-1})^{2t} \sum_{z\in\Gamma_{n_j}(x)} \exp\big(q(S_{n_j}\varphi(z) - P_x^{n_j}(\varphi))$$

$$-(T(q)+\varepsilon_1)\log|(f_x^{n_j})'(z)|)v_x^{-q}\big(f_{\hat{z}}^{-n_j}(B(f^{n_j}(\hat{z}),\xi))\big)|(f^{n_j})'(\hat{z})|^{qT'(q)-\varepsilon_2 q}$$

$$\leq (\hat{Q}\xi^{-1})^{2t} e^{q\hat{Q}_\phi} \sum_{z\in\Gamma_{n_j}(x)} (\hat{Q}\xi^{-1})^{2t} \sum_{z\in\Gamma_{n_j}(x)} \exp\big(q(S_{n_j}\varphi(z) - P_x^{n_j}(\varphi))$$

$$-(T(q)+\varepsilon_1)\log|(f_x^{n_j})'(z)|\big)$$

$$\leq (\hat{Q}\xi^{-1})^{2t} e^{q\hat{Q}_\phi} \sum_{k=1}^{a(x_{n_j})} \mathscr{L}_{\phi_{q,T(q)+\varepsilon_1},x} \mathbb{1}(w_k(x_{n_j})) \leq C(\hat{Q}\xi^{-1})^{2t} e^{q\hat{Q}_\phi} a(x_{n_j})e^{-\gamma n_j}$$

$$\leq C(\hat{Q}\xi^{-1})^{2t} e^{q\hat{Q}_\phi} a e^{-\gamma n_j}.$$

Letting $j\to\infty$ and looking also at (6.6), we thus conclude that $\mathscr{H}^t(K_x(\varepsilon_2, l)) = 0$. In virtue of (6.4) this implies that $\mathscr{H}^t(K_x(-T'(q))) = 0$. Since $\varepsilon_1 > 0$ and $\varepsilon_2 q > 0$ were arbitrary, it follows that

$$g_{\mu_x}(-T'(q)) = \mathrm{HD}(K_x(-T'(q))) \leq -qT'(q) + T(q). \tag{6.8}$$

Let us now prove the opposite inequality. For every $s \geq 1$ let s_- be the largest integer in $[0, s-1]$ such that $\theta^{s_-}(x) \in X_*$ and let s_+ be the least integer in $[s+1, +\infty)$ such that $\theta^{s_+}(x) \in X_*$. It follows from (6.2) applied with $j = l_+$ and $i = k_-$, that (5.4) is true with $s+1$ replaced by k_+, and (5.13) is true with $l-1$ replaced by l_-, that

$$\frac{\log v_{q,x}(B(y,r))}{\log r} \leq \frac{-q^* \log \hat{D}_\xi + q\big(S_{l_+}\varphi(y) - P_x^{l_+}(\varphi)\big) - T(q)\log|(f_x^{l_+})'(y)|}{\log\xi + \xi^\alpha \hat{Q} - \log|(f_x^{l_-})'(y)|}$$

and

$$\frac{\log v_{q,x}(B(y,r))}{\log r} \geq \frac{q^* \log \hat{D}_\xi + q\big(S_{k_-}\varphi(y) - P_x^{k_-}(\varphi)\big) - T(q)\log|(f_x^{k_-})'(y)|}{\log\xi - \xi^\alpha \hat{Q} - \log|(f_x^{k_+})'(y)|}.$$

Hence,

$$\limsup_{r\to 0} \frac{\log v_{q,x}(B(y,r))}{\log r}$$

$$\leq \limsup_{n\to\infty} \left(q\frac{P_x^{n_+}(\varphi) - S_{n_+}\varphi(y)}{\log|(f_x^{n_-})'(y)|}\right) + T(q)\limsup_{n\to\infty}\frac{\log|(f_x^{n_+})'(y)|}{\log|(f_x^{n_-})'(y)|} \tag{6.9}$$

and

$$\liminf_{r \to 0} \frac{\log \nu_{q,x}(B(y,r))}{\log r}$$

$$\geq \liminf_{n \to \infty} \left(q \frac{P_x^{n-}(\varphi) - S_{n_-}\varphi(y)}{\log |(f_x^{n+})'(y)|} \right) + T(q) \liminf_{n \to \infty} \frac{\log |(f_x^{n-})'(y)|}{\log |(f_x^{n+})'(y)|}. \quad (6.10)$$

Now, given $\varepsilon > 0$ and $\delta_\varepsilon > 0$ ascribed to ε according to Lemma 6.1, fix an arbitrary $\delta \in (0, \delta_\varepsilon]$. Set

$$\phi^{(1)} = \phi_{\varepsilon,\delta}^{(1)} = \phi_{(1+\delta)q, T(q)+(qT'(q)+\varepsilon)\delta} \exp(-(1+\delta)P(\phi_q))$$

and

$$\phi^{(2)} = \phi_{\varepsilon,\delta}^{(2)} = \phi_{(1-\delta)q, T(q)+(-qT'(q)+\varepsilon)\delta} \exp(-(1+\delta)P(\phi_q)).$$

Since

$$\mathscr{E}P(\phi^{(1)}) = \mathscr{E}P(\phi_{(1+\delta)q, T(q)+(qT'(q)+\varepsilon)\delta}) + (1+\delta) \int P(\phi_q) dm$$

$$= \mathscr{E}P(\phi_{(1+\delta)q, T(q)+(qT'(q)+\varepsilon)\delta})$$

and

$$\mathscr{E}P(\phi^{(2)}) = \mathscr{E}P(\phi_{(1-\delta)q, T(q)+(-qT'(q)+\varepsilon)\delta}) + (1-\delta) \int P(\phi_q) dm$$

$$= \mathscr{E}P(\phi_{(1-\delta)q, T(q)+(-qT'(q)+\varepsilon)\delta}),$$

it follows from Lemmas 6.1 and 9.5, there exists $= \kappa(q, \varepsilon, \delta) \in (0, 1)$ such that for all $k = 1, 2$, and all $n \geq 1$ sufficiently large, we have $\frac{1}{n} \log \mathscr{L}_{\phi_x^{(k)}}^n(\mathbb{1})(w) \leq \log \kappa$ for all $x \in X'_+$ and all $w \in \mathscr{J}_{\theta^n(x)}$. Equivalently,

$$\mathscr{L}_{\phi_x^{(k)}}^n(\mathbb{1})(w) \leq \kappa^n. \quad (6.11)$$

Now, for all $x \in X'_+$, all $j \geq 1$, all $1 \leq k \leq a(\theta^{n_j}(x) \leq \hat{a}$, and all $z \in f_x^{-n_j}(w_k(x_{n_j}))$, define

$$A(z) := \left\{ y \in f_z^{-n_j}(B(w_k(x_{n_j}), \xi)) : B(f^{n_j}(y), R) \subset B(w_k(x_{n_j}), \xi) \right\}.$$

Note that

$$\bigcup_{k=1}^{a(x_{n_j})} \bigcup_{z \in f_x^{-n_j}(w_k(x_{n_j}))} A(z) = \mathscr{J}(x). \quad (6.12)$$

Fix any $q \in D_T$ and set

$$\Delta_\varepsilon = \sup_{0 < \delta \le \delta_\varepsilon} \Big\{ \max\{((1 + \delta)q, T(q) + (qT'(q) + \varepsilon)\delta)^*,$$

$$((1 - \delta)q, T(q) + (-qT'(q) + \varepsilon)\delta)^*\}\Big\}.$$

Let $x \in X'_+$. Set

$$M := \exp(\hat{Q}\delta(-qT'(q) + T(q) - \varepsilon)).$$

Then, using (6.12), Lemma 2.3 (for the potential $(x, z) \mapsto \log|f'_x(z)|$, (6.2), and (6.11), we obtain

$$v_{q,x}\big(\{y \in \mathscr{J}_x : v_{q,x}\big(f_y^{-n_j}(B(f^{n_j}(y), R))\big) \ge |(f_x^{n_j})'(y)|^{-(-qT'(q)+T(q))+\varepsilon}\}\big)$$

$$= v_{q,x}\big(\{y \in \mathscr{J}_x : v_{q,x}\big(f_y^{-n_j}(B(f^{n_j}(y), R))\big)|(f_x^{n_j})'(y)|^{-qT'(q)+T(q)-\varepsilon} \ge 1\}\big)$$

$$= v_{q,x}\big(\{y \in \mathscr{J}_x : v_{q,x}^\delta\big(f_y^{-n_j}(B(f^{n_j}(y), R))\big)|(f_x^{n_j})'(y)|^{\delta(-qT'(q)+T(q)-\varepsilon)} \ge 1\}\big)$$

$$\le \int_{\mathscr{J}_x} v_{q,x}^\delta\big(f_y^{-n_j}(B(f^{n_j}(y), R))\big)|(f_x^{n_j})'(y)|^{\delta(-qT'(q)+T(q)-\varepsilon)} dv_{q,x}(y)$$

$$\le \sum_{k=1}^{a(x_{n_j})} \sum_{z \in f_x^{-n_j}(w_k(x_{n_j}))} \int_{A(z)} v_{q,x}^\delta\big(f_y^{-n_j}(B(f^{n_j}(y), R))\big)$$

$$|(f_x^{n_j})'(y)|^{\delta(-qT'(q)+T(q)-\varepsilon)} dv_{q,x}(y)$$

$$\le \sum_{k=1}^{a(x_{n_j})} \sum_{z \in f_x^{-n_j}(w_k(x_{n_j}))} v_{q,x}^\delta\big(f_z^{-n_j}(B(w_k(x_{n_j}), \xi))\big)$$

$$|(f_x^{n_j})'(z)|^{\delta(-qT'(q)+T(q)-\varepsilon)} M v_{q,x}(A(z))$$

$$\le M \sum_{k=1}^{a(x_{n_j})} \sum_{z \in f_x^{-n_j}(w_k(x_{n_j}))} v_{q,x}^\delta\big(f_z^{-n_j}(B(w_k(x_{n_j}), \xi))\big)|(f_x^{n_j})'(z)|^{\delta(-qT'(q)+T(q)-\varepsilon)}$$

$$\cdot v_{q,x}\big(f_z^{-n_j}(B(w_k(x_{n_j}), \xi))\big))$$

$$= M \sum_{k=1}^{a(x_{n_j})} \sum_{z \in f_x^{-n_j}(w_k(x_{n_j}))} v_{q,x}^{1+\delta}\big(f_z^{-n_j}(B(w_k(x_{n_j}), \xi))\big)|(f_x^{n_j})'(z)|^{\delta(-qT'(q)+T(q)-\varepsilon)}$$

$$\le M D_\xi^{\Delta_\varepsilon} \sum_{k=1}^{a(x_{n_j})} \sum_{z \in f_x^{-n_j}(w_k(x_{n_j}))} \exp\Big((1 + \delta)q\big(S_{n_j}\phi(z) - P_x^{n_j}(\phi(z)) - (1 + \delta)P_x(\phi_q^{n_j})\big)\Big)$$

$$|(f_x^{n_j})'(z)|^{-(T(q)(1+\delta)+\delta(qT'(q)-T(q)+\varepsilon))} \exp(-(1+\delta)P_x^{n_j}(\phi_q(z)))$$

$$= MD_\xi^{\Delta_\varepsilon} \sum_{k=1}^{a(x_{n_j})} \sum_{z \in f_x^{-n_j}(w_k(x_{n_j}))} \exp\left((1+\delta)q(S_{n_j}\phi(z) - P_x^{n_j}(\phi(z))) - (1+\delta)P_x(\phi_q^{n_j})\right)$$

$$\cdot |(f_x^{n_j})'(z)|^{-(T(q)+(qT'(q)+\varepsilon)\delta)} \exp(-(1+\delta)P_x^{n_j}(\phi_q(z)))$$

$$= MD_\xi^{\Delta_\varepsilon} \sum_{k=1}^{a(x_{n_j})} \mathscr{L}_{\phi_x^{n_j}}^{n_j}(\mathbb{1})(w_k(x_{n_j})) \leq MD_\xi^{\Delta_\varepsilon} a\kappa^{n_j}. \tag{6.13}$$

Therefore,

$$\sum_{j=1}^{\infty} v_{q,x}\left(\{y \in \mathscr{I}_x : \mu_{q,x}\left(f_y^{-n_j}(B(f^{n_j}(y), R))\right) \geq |(f_x^{n_j})'(y)|^{-(-qT'(q)+T(q))+\varepsilon}\}\right) < +\infty.$$

Hence, by the Borel–Cantelli Lemma, there exists a measurable set $\mathscr{I}_{1,\varepsilon,x}^q \subset \mathscr{I}_x$ such that $v_{q,x}(\mathscr{I}_{1,\varepsilon,x}^q) = 1$ and

$$\#\Big\{j \geq 1 : v_{q,x}\big(\{y \in \mathscr{I}_x : \mu_{q,x}\big(f_y^{-n_j}(B(f^{n_j}(y), R))\big)$$
$$\geq |(f_x^{n_j})'(y)|^{-(-qT'(q)+T(q))-\varepsilon}\}\big)\Big\} < \infty. \tag{6.14}$$

Arguing similarly, with the function $\phi^{(1)}$ replaced by $\phi^{(2)}$, we produce a measurable set $\mathscr{I}_{2,\varepsilon,x}^q \subset \mathscr{I}_x$ such that $v_{q,x}(\mathscr{I}_{2,\varepsilon,x}^q) = 1$ and

$$\#\Big\{j \geq 1 : v_{q,x}\big(\{y \in \mathscr{I}_x : \mu_{q,x}\big(f_y^{-n_j}(B(f^{n_j}(y), R))\big)$$
$$\leq |(f_x^{n_j})'(y)|^{-(-qT'(q)+T(q))+\varepsilon}\}\big)\Big\} < \infty. \tag{6.15}$$

Set

$$\mathscr{I}_x^q = \bigcap_{n=1}^{\infty} \mathscr{I}_{1,1/n,x}^q \cap \mathscr{I}_{2,1/n,x}^q.$$

Then $v_{q,x}(\mathscr{I}_x^q) = 1$ and, it follows from (6.13) and (6.1), that for all $y \in \mathscr{I}_x^q$, we have

$$\lim_{j \to \infty} \frac{q(P_x^{n_j}(\varphi) - S_{n_j}\varphi(y))}{\log|(f_x^{n_j})'(y)|} = -qT'(q).$$

Since $\lim_{n \to \infty} \frac{n_-}{n_+} = 1$, it thus follows from (6.9) and (6.10) that

$$d_{v_{q,x}}(y) = -qT'(q) + T(q), \tag{6.16}$$

and (recall that $v_{1,x} = v_x$ and $T(1) = 0$)

$$\lim_{r \to 0} \frac{\log v_x(B(x,r))}{\log r} = -T'(q)$$

for all $y \in \mathscr{J}_x^q$. As the latter formula implies that $\mathscr{J}_x^q \subset K(-T'(q))$, and as $v_{q,x}(\mathscr{J}_x^q) = 1$, applying (6.16), we get that

$$g_{\mu_x}(-T'(q)) = \mathrm{HD}(K_x(-T'(q))) \geq \mathrm{HD}(\mathscr{J}_x^q)) = -qT'(q) + T(q).$$

Combining this formula with (6.8) completes the proof.   □

As an immediate consequence of this proposition we get the following theorem.

Theorem 6.3 *Suppose that* $f(x,z) = (\theta(x), f_x(z))$ *is a conformal random expanding map. Then the Legendre conjugate,* $g: \mathrm{Range}(-T') \to [0, +\infty)$, *to the temperature function* $\mathbb{R} \ni q \mapsto T(q)$ *is differentiable everywhere except a countable set of points, call it* D_T^*, *and there exists a measurable set* $X_{ma} \subset X$ *with* $m(X_{ma}) = 1$ *such that for every* $\alpha \in D_T^*$ *and every* $x \in X_{ma}$, *we have*

$$g_{\mu_x}(\alpha) = g(\alpha).$$

6.3 Analyticity of the Multifractal Spectrum for Uniformly Expanding Random Maps

Now, as in Chap. 9.4, we assume that we deal with a conformal uniform random expanding map. In particular, the essential infimum of γ_x is larger than some $\gamma > 1$ and functions H_x, $n_\xi(x)$, $j(x)$ are finite. In addition, we have that there exist constants L and $c > 0$ such that

$$S_n \varphi_x(y) \leq -nc + L \tag{6.17}$$

for every $y \in \mathscr{J}_x$ and n and $\mathscr{E}P(\varphi) = 0$. With these assumptions we can get the following property of the function T.

Proposition 6.4 *Suppose that* $f: \mathscr{J} \to \mathscr{J}$ *is a conformal uniformly random expanding map. Then the temperature function* T *is real-analytic and for every* q, *we have*

$$T'(q) = \frac{\int_{\mathscr{J}} \varphi d\mu_q}{\int_{\mathscr{J}} \log |f'| d\mu_q} < 0. \tag{6.18}$$

Proof. The potentials

$$\varphi_{q,x,t}(y) := q(\varphi_x(y) - P_x(\varphi)) - t \log |f_x'(y)|$$

extend by the same formula to holomorphic functions $\mathbb{C} \times \mathbb{C} \ni (q, t) \mapsto \varphi_{q,x,t}(y)$. Since these functions are in fact linear, we see that the assumptions of Theorem 9.17 are satisfied, and therefore the function $\mathbb{R} \times \mathbb{R} \ni (q, t) \mapsto \mathscr{E}P(q, t)$ is real-analytic. Since $|f'_x(y)| > 0$, in virtue of Proposition 9.18 we obtain that

$$\frac{\partial \mathscr{E}P(q, t)}{dt} = -\int_{\mathscr{J}} \log|f'_x| d\mu_{q,x,t} dm(x) < 0. \tag{6.19}$$

Hence, we can apply the Implicit Function Theorem to conclude that the temperature function $\mathbb{R} \ni q \mapsto T(q) \in \mathbb{R}$, satisfying the equation,

$$\mathscr{E}P(q, T(q)) = 0,$$

is real-analytic. Hence,

$$0 = \frac{d\mathscr{E}P(\varphi_q)}{dq} = \frac{\partial \mathscr{E}P(q, t)}{\partial q}\bigg|_{t=T(q)} + \frac{\partial \mathscr{E}P(q, t)}{\partial t}\bigg|_{t=T(q)} T'(q).$$

Then

$$T'(q) = -\frac{\frac{\partial \mathscr{E}P(q,t)}{\partial q}\big|_{t=T(q)}}{\frac{\partial \mathscr{E}P(q,t)}{\partial t}\big|_{t=T(q)}} = -\frac{\int_{\mathscr{J}} (\varphi_x - P_x) d\mu_{q,x} dm(x)}{\int_{\mathscr{J}} -\log|f'_x| d\mu_{q,x} dm(x)}$$

$$= \frac{\int_{\mathscr{J}} \varphi_x d\mu_{q,x} dm(x) - \int_X P_x dm(x)}{\int_{\mathscr{J}} \log|f'_x| d\mu_{q,x} dm(x)} = \frac{\int_{\mathscr{J}} \varphi d\mu_q}{\int_{\mathscr{J}} \log|f'| d\mu_q}.$$

So, we obtain (6.18). It follows, in particular, that

$$T'(q) < 0, \tag{6.20}$$

since by (6.17), the integral $\int_{\mathscr{J}} \varphi d\mu_q$ is negative.   □

Combining this proposition with Proposition 6.2 we get the following result which concludes this section.

Theorem 6.5 *Suppose that $f : \mathscr{J} \to \mathscr{J}$ is a conformal uniformly random expanding map. Then the Legendre conjugate, $g : \text{Range}(-T') \to [0, +\infty)$, to the temperature function $\mathbb{R} \ni q \mapsto T(q)$ is real-analytic, and there exists a measurable set $X_{ma} \subset X$ with $m(X_{ma}) = 1$ such that for every $\alpha \in \text{Range}(-T')$ and every $x \in X_{ma}$, we have*

$$g_{\mu_x}(\alpha) = g(\alpha).$$

Chapter 7
Expanding in the Mean

In this chapter we show that the main achievements of this manuscript, including thermodynamical formalism, Bowen's formula and multifractal analysis, also hold for a class of random maps satisfying an allegedly weaker expanding condition

$$\int \log \gamma_x dm(x) > 0.$$

We start with a precise definition of this class. Then we explain how this case can be reduced to random expanding maps by looking at an appropriate induced map. The picture is completed by providing and discussing a concrete map that is not expanding but expanding in the mean.

7.1 Definition of Maps Expanding in the Mean

Let $T : \mathscr{J} \to \mathscr{J}$ be a skew-product map as defined in Sect. 2.2 satisfying the properties of *Measurability of the Degree* and *Topological Exactness*. Such a random map is called *expanding in the mean*, if for some $\xi > 0$ and some measurable function $X \ni x \mapsto \gamma_x \in \mathbb{R}_+$ with

$$\int \log \gamma_x dm(x) > 0,$$

we have that all inverse branches of every T_x^n are well defined on balls of radii ξ and are $(\gamma_x^n)^{-1}$-Lipschitz continuous. More precisely, for every $y = (x, z) \in \mathscr{J}$ and every $n \in \mathbb{N}$, there exists

$$T_y^{-n} : B_{\theta^n(x)}(T^n(y), \xi) \to \mathscr{J}_x$$

V. Mayer et al., *Distance Expanding Random Mappings, Thermodynamical Formalism, Gibbs Measures and Fractal Geometry*, Lecture Notes in Mathematics 2036, DOI 10.1007/978-3-642-23650-1_7, © Springer-Verlag Berlin Heidelberg 2011

such that

1. $T^n \circ T_y^{-n} = \mathrm{Id}|_{B_{\theta^n(x)}(T^n(y),\xi)}$ and $T_y^{-n}(T_x^n(z)) = z$,
2. $\varrho(T_y^{-n}(z_1), T_y^{-n}(z_2)) \leq (\gamma_x^n)^{-1} \varrho(z_1, z_2)$ for all $z_1, z_2 \in B_{\theta^n(x)}(T^n(y), \xi)$.

7.2 Associated Induced Map

In this section we show how the expanding in the mean maps can be reduced to our
setting from Sect. 2.3.

Let $T : \mathscr{J} \to \mathscr{J}$ be an expanding in the mean random map. To this map and to
a set $A \subset X$ of positive measure we associate an induced map $\overline{T}$ in the following
way. Let τ_A be the first return map to the set A, that is

$$\tau_A(x) = \min\{n \geq 1 : \theta^n(x) \in A\}.$$

Define also

$$\theta_A(x) := \theta^{\tau_A(x)}(x) \quad \text{and} \quad \gamma_{A,x} := \prod_{j=0}^{\tau_A(x)-1} \gamma_{\theta^j(x)}.$$

Then the *induced map* $\overline{T}$ is the random map over $(A, \mathscr{B}, m_A)$ defined by

$$\overline{T}_x = T_x^{\tau_A(x)} \quad \text{for a.e. } x \in A.$$

The following lemma show that the set A can be chosen such that $\overline{T}$ is an
expanding random map.

Lemma 7.1 *There exists a measurable set $A \subset X$ with $m(A) > 0$ such that*

$$\gamma_{A,x} > 1 \quad \text{for all } x \in A.$$

Proof. First, define inductively

$$A_1 := \{x : \log \gamma_x > 0\}$$

and, for $k \geq 1$,

$$A_{k+1} := \{x \in A_k : \log \gamma_{A_k,x} > 0\}.$$

Since

$$0 < \int_X \log \gamma_x dm(x) = \int_{A_1} \log \gamma_{A_1,x} dm(x) = \int_{A_k} \log \gamma_{A_k,x} dm(x),$$

we have that $m(A_k) > 0$ for all $k \geq 1$. Obviously, the sequence $(A_k)_{k=1}^{\infty}$ is decreasing. Let

$$A = \bigcap_{k=1}^{\infty} A_k \quad \text{and} \quad E = X \setminus A.$$

Notice that the points $x \in E$ have the property that $\log \gamma_x^n \leq 0$ for some $n \geq 1$.

Claim: $m(A) > 0$.

If on the contrary $m(A) = \lim_{k \to \infty} m(A_k) = 0$, then $m(E) = 1$. Since the measure m is θ-invariant, we have that $m(E_{\infty}) = 1$ where

$$E_{\infty} = \bigcap_{n=0}^{\infty} \theta^{-n}(E).$$

For $x \in E_{\infty}$ we have that $\log \gamma_x^n \leq 0$ for infinitely many $n \geq 1$. This contradicts Birkhoff's Ergodic Theorem since, by hypothesis, $\int \log \gamma_x > 0$. Therefore the set A has positive measure.

Since $m(A) > 0$, τ_A is almost surely finite. Now let $x \in A$. Then, for every point $\theta^j(x)$, $j = 1, \ldots, \tau_A(x) - 1$, we can find $k(j)$ such that $\theta^j(x) \in X \setminus A_{k(j)}$. Put

$$K(x) = \max\{k(j) : j = 1, \ldots, \tau_A(x) - 1\} + 1.$$

Hence x and $\theta_A(x)$ are in $A_{K(x)}$ and $\theta^j(x) \notin A_{K(x)}$ for $j = 1, \ldots, \tau_A(x) - 1$. Hence $\tau_A(x) = \tau_{A_{K(x)}}(x)$, and therefore

$$\gamma_{A,x} = \gamma_{A_{K(x)},x} > 1. \qquad \Box$$

In the following $A \subset X$ will be some set coming from Lemma 7.1 and $\overline{T} = T^{\tau_A}$ the associated induced map. For this map, we have to consider the following appropriated class of Hölder potentials. First, to every $y = (x, z)$ we associate the neighborhood

$$U(z) = \bigcup_{n=0}^{\infty} T_y^{-n}\big(B_{\theta^n(x)}(T^n(y), \xi)\big) \subset \mathscr{J}_x.$$

Fix $\alpha \in (0, 1]$. As in Sect. 2.7 a function $\varphi \in \mathscr{C}^1(\mathscr{J})$ is called *Hölder continuous with an exponent* α provided that there exists a measurable function $H : X \to [1, +\infty)$, $x \mapsto H_x$, such that

$$\int_X \log H_x \, dm(x) < \infty \qquad (7.1)$$

and

$$v_\alpha(\varphi_x) \le H_x \quad \text{for a.e. } x \in X.$$

The subtlety here is that the infimum in the definition (2.11) of v_α is now taken over all $z_1, z_2 \in \mathscr{J}_x$ with $z_1, z_2 \in U(z)$, $z \in \mathscr{J}_x$. For example, any function, which is H_x-Hölder on the entire set $\mathscr{J}_x$ is fine.

Let T be an expanding in the mean random map and φ a Hölder potential according to the definition above. Having associated to T the induced map $\overline{T}$, one naturally has to replace the potential φ by the *induced potential*

$$\overline{\varphi}_x(z) = \sum_{j=0}^{\tau_A(x)-1} \varphi_{\theta^j(x)}(T_x^j(z)).$$

Although, it is not clear if the potential $\overline{\varphi}$ satisfies the condition (7.1), the choice of the neighborhoods $U(z)$ and the definition of Hölder potentials make that Lemma 2.3 still holds. This gives us an important control of the distortion which is what is needed in the rest of the paper rather than the condition (7.1) leading to it. The hypothesis (7.1) is only used in the proof of Lemma 2.3.

7.3   Back to the Original System

In this section we explain how to get the Thermodynamic Formalism for the original system.

With the preceding notations, for the expanding induced map $\overline{T}$ the Thermo-dynamical Formalism of Chap. 3 and, in particular, the Theorems 3.1 and 3.2 do apply. We denote by $\overline{v}_x$, $\overline{\mu}_x$ and $\overline{q}_x$, $x \in A$, the resulting conformal and invariant measures and the invariant density, respectively, for $\overline{T}$. We now explain how the corresponding objects can be recovered for the original map T. Notice that this is possible since we only induced in the base system.

First, we consider the case of the conformal measures. Let $\overline{v}_x$, $x \in A$ be the measure such that

$$\overline{\mathscr{L}}_x^* \overline{v}_{\theta_A(x)} = \overline{\lambda}_x \overline{v}_x.$$

If $x \in A$ we put $v_x = \overline{v}_x$. If $x \notin A$, then by ergodicity of θ, almost surely there exists $k \in \mathbb{N}$, such that $\theta^k(x) \in A$ and $\theta^j(x) \notin A$ for $j = 0, \ldots, k-1$. Then we put

$$v_x = \frac{(\mathscr{L}_x^k)^* v_{\theta^k(x)}}{\mathscr{L}_x^k(\mathbb{1})}. \tag{7.2}$$

Therefore, the family $\{v_x\}_{x \in X'}$ is a family of probability measures well defined for X in a subset X' of X with full measure. This family of measures has the conformality property

$$\mathscr{L}_x^* v_{\theta^k(x)} = \lambda_x v_x,$$

where $\lambda_x = v_{\theta(x)}(\mathscr{L}_x \mathbb{1})$, $x \in X'$. Notice also that $\mathscr{E}P(\varphi) = \mathscr{E}P(\overline{\varphi})$.

Similarly, from the family $\{\overline{\mu}_x\}_{x \in A}$ of $\overline{T}$-invariant measures one can recover a family $\{\mu_x\}_{x \in X}$ of invariant measures for the original map T. Indeed, for $x \in A$ and $j = 0, \ldots, \tau_A(x) - 1$ it suffices to put

$$\mu_{\theta^j(x)} = \overline{\mu}_x \circ T_x^{-j}.$$

Then, with $q_{\theta^j(x)} = \mathcal{L}_x^j(\overline{q}_x)$, we have that

$$d\mu_{\theta^j(x)} = q_{\theta^j(x)} d\nu_{\theta^j(x)}.$$

Hence Theorems 3.1 and 3.2 among with all statistical consequences hold for the original map. Moreover, since $\mathcal{E}P(\overline{\varphi}_t) = \mathcal{E}P(\varphi_t)$ their zeros coincide and consequently Bowen's Formula and the Multifractal Analysis are also true for conformal expanding in the mean random maps.

7.4 An Example

Here is an example of an expanding in the mean random system. Define

$$f_0(x) = \begin{cases} \frac{1}{2}x + \frac{15}{2}x^2 & \text{if } x \in [0, 1/3] \\ 16x - 15 & \text{if } x \in [15/16, 1] \end{cases}$$

and

$$f_1(x) = 16x \,(\text{mod}\, 1) \text{ for } x \in [0, 1/16] \cup [15/16, 1].$$

Let $X = \{0, 1\}^{\mathbb{Z}}$, θ be the shift transformation and m be the standard Bernoulli measure. For $x = (\ldots, x_{-1}, x_0, x_1, \ldots) \in X$ define

$$f_x = f_{x_0}, \; f_x^n = f_{\theta^{n-1}(x)} \circ f_{\theta^{n-2}(x)} \circ \ldots \circ f_x$$

and

$$\mathcal{J}_x = \bigcap_{n=0}^{\infty} (f_x^n)^{-1}([0, 1]).$$

For this map, $\gamma_0 = 1/2$ and $\gamma_1 = 16$ are the best expanding constants that one can take. With these constants we have

$$\int \log \gamma_x \, dm(x) > 0.$$

Therefore, the map is expanding in the mean but not expanding.

Note that the size of each component of $f_x^{-n}([0, 1])$ is bounded by

$$a_n = 16^{-n_1}(1/2)^{-n_0}, \tag{7.3}$$

where $n_i := \#\{j = 0, \ldots, n - 1 : x_j = i\}, i = 0, 1$. Since

$$\lim_{n \to \infty} \frac{n_0}{n} = \lim_{n \to \infty} \frac{n_1}{n} = \frac{1}{2}$$

almost surely, we have that $\lim_{n \to \infty} a_n = 0$. Hence, for almost every $x \in X$, $\mathscr{J}_x$ is a Cantor set. Moreover, by (7.3), almost surely we have, that,

$$\mathscr{E}P(t) \leq \lim_{n \to \infty} \frac{1}{n} \log 2^n 16^{-n_1 t} (1/2)^{-n_0 t}$$

$$\leq \log 2 - t \Big(\lim_{n \to \infty} \frac{n_1}{n} \log 16 - \frac{n_0}{n} \log 2 \Big) = \log 2 \left(1 - \frac{3}{2}t \right).$$

Therefore, by Bowen's Formula, the Hausdorff dimension of almost every fiber $\mathscr{J}_x$ is smaller than or equal to $2/3$. Notice however that for some choices of $x \in X$ the fiber $\mathscr{J}_x$ contains open intervals.

Chapter 8
Classical Expanding Random Systems

Having treated a very general situation up to here, we now focus on more concrete random repellers and, in the next section, random maps that have been considered by Denker and Gordin. The Cantor example of Chap. 5.3 and random perturbations of hyperbolic rational functions like the examples considered by Brück and Büger are typical random maps that we consider now. We classify them into quasi-deterministic and essential systems and analyze then their fractal geometric properties. Here as a consequence of the techniques we have developed, we positively answer the question of Brück and Büger (see [9] and Question 5.4 in [8]) of whether the Hausdorff dimension of almost all (most) naturally defined random Julia sets is strictly larger than 1. We also show that in this same setting the Hausdorff dimension of almost all Julia sets is strictly less than 2.

8.1 Definition of Classical Expanding Random Systems

Let (Y, ρ) be a compact metric space normalized by $diam(Y) = 1$ and let $U \subset Y$. A *repeller over U* will be a continuous open and surjective map $T : V_T \to U$ where $\overline{V_T}$, the closure of the domain of T, is a subset of U. Let $\gamma > 1$ and consider

$$\mathcal{R} = \mathcal{R}(U, \gamma) = \{T : V_T \to U \quad \gamma\text{-expanding repeller over } U\}.$$

Concerning the randomness we will consider classical independently and identically distributed (i.i.d.) choices. More precisely, we suppose the repellers

$$T_{x_0}, T_{x_1}, ..., T_{x_n}, ... \tag{8.1}$$

are chosen i.i.d. with respect to some arbitrary probability space $(I, \mathcal{F}_0, m_0)$. This gives rise to a *random repeller* $T_{x_0}^n = T_{x_{n-1}} \circ ... \circ T_{x_0}, n \geq 1$. The natural associated Julia set is

V. Mayer et al., *Distance Expanding Random Mappings, Thermodynamical Formalism,* *Gibbs Measures and Fractal Geometry*, Lecture Notes in Mathematics 2036, DOI 10.1007/978-3-642-23650-1_8, © Springer-Verlag Berlin Heidelberg 2011

$$\mathscr{J}_x = \bigcap_{n \geq 1} T_{x_0}^{-n}(U) \quad \text{where} \quad x = (x_0, x_1, \ldots).$$

Notice that compactness of Y together with the expanding assumption, we recall that γ-expanding means that the distance of all points z_1, z_2 with $\rho(z_1, z_2) \leq \eta_T$ is expanded by the factor γ, implies that $\mathscr{J}_x$ is compact and also that the maps $T \in \mathscr{R}$ are of bounded degree. A random repeller is therefore the most classical form of a uniformly expanding random system.

The link with the setting of the preceding sections goes via natural extension. Set $X = I^{\mathbb{Z}}$, take the Bernoulli measure $m = m_0^{\mathbb{Z}}$ and let the ergodic invariant map θ be the shift map $\sigma : I^{\mathbb{Z}} \to I^{\mathbb{Z}}$. If $\pi : X \to I$ is the projection on the 0th coordinate and if $x \mapsto T_x$ is a map from I to $\mathscr{R}$ then the repeller (8.1) is given by the skew-product

$$T(x, z) = \big(\sigma(x), T_{\pi(x)}(z)\big), \quad (x, z) \in \mathscr{J} = \bigcup_{x \in X} \{x\} \times \mathscr{J}_x. \tag{8.2}$$

The particularity of such a map is that the mappings T_x do only depend on the 0th coordinate. It is natural to make the same assumption for the potentials, i.e. $\varphi_x = \varphi_{\pi(x)}$. We furthermore consider the following continuity assumptions:

(T0) I is a bounded metric space.

(T1) $(x, z) \mapsto T_x^{-1}(z)$ is continuous from $\mathscr{J}$ to $\mathscr{K}(U)$, the space of all non-empty compact subsets of U equipped with the Hausdorff distance.

(T2) For every $z \in U$, the map $x \mapsto \varphi_x(z)$ is continuous.

A *classical expanding random system* is a random repeller together with a potential depending only on the 0th-coordinate such that the conditions (T0), (T1) and (T2) hold.

Example 8.1 *Suppose V, U are open subsets of $\mathbb{C}$ with V compactly contained in U and consider the set $\mathscr{R}(V, U)$ of all holomorphic repellers $T : V_T \to U$ having uniformly bounded degree and a domain $V_T \subset V$. This space has natural topologies, for example the one induced by the distance*

$$\rho(T_1, T_2) = d_H\big(V_{T_1}, V_{T_2}\big) + \|(T_1 - T_2)_{|V_{T_1} \cap V_{T_2}}\|_\infty,$$

where d_H denotes the Hausdorff metric. Taking then geometric potentials $-t \log |T'|$ we get one of the most natural example of classical expanding random system.

Proposition 8.2 *The pressure function $x \mapsto P_x(\varphi)$ of a classical expanding random system is continuous.*

Proof. We have to show that $x \mapsto \lambda_x$ is continuous and since $\mathscr{L}_x^n \mathbb{1}(y) / \mathscr{L}_{x_1}^{n-1} \mathbb{1}(y)$ converges uniformly to λ_x for every $y \in U$ (see Lemma 3.32) it suffices to show that $x \mapsto \mathscr{L}_x^n \mathbb{1}(y)$ does depend continuously on $x \in X$. In order to do so, we first show that condition (T1) implies continuity of the function $(x, y) \mapsto \# T_x^{-1}(y)$.

Let $(x, y) \in X \times U$ and fix $0 < \xi' < \xi$ such that $B(w_1, \xi') \cap B(w_2, \xi') = \emptyset$ for all disjoint $w_1, w_2 \in T_x^{-1}(y)$. From (T1) follows that there exists $\delta > 0$ such that

$$d_H(T_x^{-1}(y), T_{x'}^{-1}(y')) \leq \frac{\xi}{2}, \quad \text{whenever } \varrho((x, y), (x', y')) \leq \delta.$$

But this implies that for every $w \in T_x^{-1}(y)$ there exists at least one preimage $w' \in T_{x'}^{-1}(y') \cap B(w, \xi')$. Consequently, $\#T_{x'}^{-1}(y') \geq \#T_x^{-1}(y)$. Equality follows since $T_{x'}$ is injective on every ball of radius ξ', a consequence of the expanding condition.

Let $x \in X$, let W be a neighborhood of x and let $y \in U$. From what was proved before we have that for every $w \in T_x^{-1}(y)$, there exists a continuous function $x' \mapsto z_w(x')$ defined on W such that $T_{x'}(z_w(x')) = y$, $z_w(x) = w$ and

$$T_{x'}^{-1}(y) = \{z_w(x') : w \in T_x^{-1}(y)\}.$$

The proposition follows now from the continuity of φ_x, i.e. from (T2). □

We say that a function $g : I^{\mathbb{Z}} \to \mathbb{R}$ is past independent if $g(\omega) = g(\tau)$ for any $\omega, \tau \in I^{\mathbb{Z}}$ with $\omega|_0^{\infty} = \tau|_0^{\infty}$. Fix $\kappa \in (0, 1)$ and for every function $g : I^{\mathbb{Z}} \to \mathbb{R}$ set

$$v_\kappa(g) = \sup_{n \geq 0} \{v_{\kappa,n}(g)\},$$

where

$$v_{\kappa,n}(g) = \kappa^{-n} \sup\{|g(\omega) - g(\tau)| : \omega|_0^n = \tau|_0^n\}.$$

Denote by H_κ the space of all bounded Borel measurable functions $g : I^{\mathbb{Z}} \to \mathbb{R}$ for which $v_\kappa(g) < +\infty$. Note that all functions in H_κ are past independent. Let $\mathbb{Z}_-$ be the set of negative integers. If I is a metrizable space and d is a bounded metric on I, then the formula

$$d_+(\omega, \tau) = \sum_{n=0}^{\infty} 2^{-n} d(\omega_n, \tau_n)$$

defines a pseudo-metric on $I^{\mathbb{Z}}$, and for every $\tau \in I^{\mathbb{Z}}$, the pseudo-metric d_+ restricted to $\{\tau\} \times \mathbb{N}$, becomes a metric which induces the product (Tychonoff) topology on $\{\tau\} \times \mathbb{N}$.

Theorem 8.3 *Suppose that* $T : \mathscr{J} \to \mathscr{J}$ *and* $\phi : \mathscr{J} \to \mathbb{R}$ *form a classical expanding random system. Let* $\lambda : I^{\mathbb{Z}} \to (0, +\infty)$ *be the corresponding function coming from Theorem 3.1. Then both functions* λ *and* $P(\phi)$ *belong to* H_κ *with some* $\kappa \in (0, 1)$, *and both are continuous with respect to the pseudo-metric* d_+.

Proof. Let $y \in U$ be any point. Fix $n \geq 0$ and $\omega, \tau \in I^{\mathbb{Z}}$ with $\omega|_0^n = \tau|_0^n$. By Lemma 3.32, we have

$$\left| \frac{\mathscr{L}_\omega^{n+1} \mathbb{1}(y)}{\mathscr{L}_{\sigma(\omega)}^n \mathbb{1}(y)} - \lambda_\omega \right| \leq A\kappa^n \quad \text{and} \quad \left| \frac{\mathscr{L}_\tau^{n+1} \mathbb{1}(y)}{\mathscr{L}_{\sigma(\tau)}^n \mathbb{1}(y)} - \lambda_\tau \right| \leq A\kappa^n$$

with some constants $A > 0$ and $\kappa \in (0, 1)$. Since, by our assumptions, $\mathscr{L}_\tau^{n+1} \mathbb{1}(y) = \mathscr{L}_\omega^{n+1} \mathbb{1}(y)$ and $\mathscr{L}_{\sigma(\omega)}^n \mathbb{1}(y) = \mathscr{L}_{\sigma(\tau)}^n \mathbb{1}(y)$, we conclude that $|\lambda_\omega - \lambda_\tau| \leq 2A\kappa^n$. So,

$$v_\kappa(\lambda) \leq 2A.$$

Since, by Proposition 8.2, the function $\lambda : I^\mathbb{Z} \to (0, +\infty)$ is continuous, it is therefore bounded above and separated from zero. In conclusion, both functions λ and $P(\phi)$ belong to H_κ with some $\kappa \in (0, 1)$, and both are continuous with respect to the pseudo-metric d_+. $\square$

Corollary 8.4 *Suppose that $T : \mathscr{J} \to \mathscr{J}$ and $\phi : I^\mathbb{Z} \to \mathbb{R}$ form a classical expanding random system. Then the number (asymptotic variance of $P(\phi)$)*

$$\sigma^2(P(\phi)) = \lim_{n \to \infty} \frac{1}{n} \int \left(S_n(P(\phi)) - n\mathscr{E}P(\phi) \right)^2 dm \geq 0$$

exists, and the Law of Iterated Logarithm holds, i.e. m-a.e we have

$$-\sqrt{2\sigma^2(P(\phi))} = \liminf_{n \to \infty} \frac{P_x^n - n\mathscr{E}P(\phi)}{\sqrt{n \log \log n}} \leq \limsup_{n \to \infty} \frac{P_x^n(\phi) - n\mathscr{E}P(\phi)}{\sqrt{n \log \log n}} = \sqrt{2\sigma^2(P(\phi))}.$$

Proof. Let $\pi : I^\mathbb{Z} \to I$ be the canonical projection onto the 0th coordinate and let $\mathscr{G} = \pi^{-1}(\mathscr{B})$, where $\mathscr{B}$ is the σ-algebra of Borel sets of I. We want to apply Theorem 1.11.1 from [24]. Condition (1.11.6) is satisfied with the function ϕ (object being here as in Theorem 1.11.1 and by no means our potential!) identically equal to zero since $|m(A \cap B) - m(A)m(B)| = 0$ for every $A \in \mathscr{G}_0^m :=$ $\mathscr{G} \cap \sigma^{-1}(\mathscr{G}) \cap \ldots \sigma^{-m}(\mathscr{G})$ and $B \in \mathscr{G}_n^\infty = \bigcap_{j=n}^{+\infty} \sigma^{-j}(\mathscr{G})$, whenever $n > m$. The integral $\int |P(\phi)|^{2+\delta} dm$ is finite (for every $\delta > 0$) since, by Theorem 8.3, the pressure function $P(\phi)$ is bounded. This then implies that for all $n \geq 1$, $|P(\phi)(\omega) - \mathscr{E}(P(\phi)|\mathscr{G}_0^n)(\omega)| \leq v_\kappa(P(\phi))\kappa^n$, where $v_\kappa(P(\phi)) < +\infty$. Therefore,

$$\int |P(\phi) - \mathscr{E}(P(\phi)|\mathscr{G}_0^n)| dm \leq v_\kappa(P(\phi))\kappa^n,$$

whence condition (1.11.7) from [24] holds. Finally, $P(\phi)$ is $\mathscr{G}_0^\infty$-measurable, since $P(\phi)$ belonging to H_κ is past independent. We have thus checked all the assumptions of Theorem 1.11.1 from [24] and, its application yields the existence of the asymptotic variance of $P(\phi)$ and the required Law of Iterated Logarithm to hold. $\square$

Proposition 8.5 *Let $g \in \mathscr{H}_\kappa$. Then $\sigma^2(g) = 0$ if and only if there exists $u \in C((\operatorname{supp}(m_0))^\mathbb{Z})$ such that $g - m(g) = u - u \circ \sigma$ holds throughout $(\operatorname{supp}(m_0))^\mathbb{Z}$.*

Proof. Denote the topological support of m_0 by S. The implication that the cohomology equation implies vanishing of σ^2 is obvious. In order to prove the other implication, assume without loss of generality that $m(g) = 0$. Because of Theorem 2.51 from [16] there exists $u \in L_2(m)$ independent of the past (as so is g) such that

$$g = u - u \circ \sigma \tag{8.3}$$

in the space $L_2(m)$. Our goal now is to show that u has a continuous version and (8.3) holds at all points of $S^{\mathbb{Z}}$. In view of Lusin's Theorem there exists a compact set $K \subset S^{\mathbb{Z}}$ such that $m(K) > 1/2$ and the function $u|_K$ is continuous. So, in view of Birkhoff's Ergodic Theorem there exists a Borel set $B \subset S^{\mathbb{Z}}$ such that $m(B) = 1$, for every $\omega \in B$, $\sigma^{-n}(\omega) \in K$ with asymptotic frequency $> 1/2$, u is well defined on $\bigcup_{n=-\infty}^{+\infty} \sigma^{-n}(B)$, and (8.3) holds on $\bigcup_{n=-\infty}^{+\infty} \sigma^{-n}(B)$. Let $\mathbb{Z}_- = \{-1, -2, \ldots\}$ and let $\{m_\tau\}_{\tau \in I^{\mathbb{Z}_-}}$ be the canonical system of conditional measures for the partition $\{\{\tau\} \times I^{\mathbb{N}}\}_{\tau \in I^{\mathbb{Z}_-}}$ with respect to the measure m. Clearly, each measure m_τ, projected to $I^{\mathbb{N}}$, coincides with m_+. Since $m(B) = 1$, there exists a Borel set $F \subset S^{\mathbb{Z}_-}$ such that $m_-(F) = 1$ and $m_\tau(B \cap (\{\tau\} \times I^{\mathbb{N}})) = 1$ for all $\tau \in F$, where m_- is the infinite product measure on $S^{\mathbb{Z}_-}$. Fix $\tau \in F$ and set $Z = p_{\mathbb{N}}(B \cap (\{\tau\} \times I^{\mathbb{N}}))$, where $p_{\mathbb{N}} : I^{\mathbb{Z}} \to I^{\mathbb{N}}$ is the natural projection from $I^{\mathbb{Z}}$ to $I^{\mathbb{N}}$. The property that $m_\tau(B \cap (\{\tau\} \times I^{\mathbb{N}})) = 1$ implies that $\overline{Z} = S^{\mathbb{N}}$. Now, it immediately follows from the definitions of Z and B that for all $x, y \in Z$ there exists an increasing sequence $(n_k)_{k=1}^{\infty}$ of positive integers such that $\sigma^{-n_k}(\tau x)$, $\sigma^{-n_k}(\tau y) \in K$ for all $k \geq 1$. For every $0 < q \leq n_k$ we have from (8.3) that

$$\sum_{j=0}^{n_k-q} \left(g(\sigma^j(\sigma^{-n_k}(\tau y))) - g(\sigma^j(\sigma^{-n_k}(\tau x))) \right)$$

$$+ \sum_{j=n_k-q+1}^{n_k} \left(g(\sigma^j(\sigma^{-n_k}(\tau y))) - g(\sigma^j(\sigma^{-n_k}(\tau x))) \right)$$

$$= (u(\sigma^{-n_k}(\tau y)) - u(\sigma^{-n_k}(\tau x)) + (u(\tau x) - u(\tau y)).$$

Since $g \in H_\kappa$, we have

$$\sum_{j=0}^{n_k-q} \left(g(\sigma^j(\sigma^{-n_k}(\tau y))) - g(\sigma^j(\sigma^{-n_k}(\tau y))) \right)$$

$$\leq \sum_{j=0}^{n_k-q} |g(\sigma^j(\sigma^{-n_k}(\tau y))) - g(\sigma^j(\sigma^{-n_k}(\tau y)))|$$

$$\leq \sum_{j=0}^{n_k-q} v_\kappa(g)\kappa^{n_k-j} \leq v_\kappa(g)(1-\kappa)^{-1}\kappa^q.$$

Now, fix $\varepsilon > 0$. Take $q \geq 1$ so large that $v_\kappa(g)(1-\kappa)^{-1}\kappa^q < \varepsilon/2$. Since the function $g : I^{\mathbb{Z}} \to \mathbb{R}$ is uniformly continuous with respect to the pseudo-metric d, there exists $\delta > 0$ such that $|g(b) - g(a)| < \frac{\varepsilon}{2q}$ whenever $d(a, b) < \delta$. Assume that $d(x, y) < \delta$ (so $d(\sigma^{-i}(\tau x), \sigma^{-i}(\tau y)) < \delta$ for all $i \geq 0$). It follows now that for every $k \geq 1$ we have

$$|u(\tau x) - u(\tau y)| \le v_\kappa(g)(1 - \kappa)^{-1}\kappa^q + q\frac{\varepsilon}{2q} + |u(\sigma^{-n_k}(\tau y)) - u(\sigma^{-n_k}(\tau x))|$$

$$\le \frac{\varepsilon}{2} + \frac{\varepsilon}{2} + |u(\sigma^{-n_k}(\tau y)) - u(\sigma^{-n_k}(\tau x))|$$

$$= \varepsilon + \frac{\varepsilon}{2} + |u(\sigma^{-n_k}(\tau y)) - u(\sigma^{-n_k}(\tau x))|.$$

Since $\sigma^{-n_k}(\tau x), \sigma^{-n_k}(\tau y) \in K$ for all $k \ge 1$, since $\lim_{k\to\infty} d(\sigma^{-n_k}(\tau x), \sigma^{-n_k}(\tau y)) = 0$, and since the function u, restricted to K, is uniformly continuous, we conclude that

$$\lim_{k\to\infty} |u(\sigma^{-n_k}(\tau y)) - u(\sigma^{-n_k}(\tau x))| = 0.$$

We therefore get that $|u(\tau x) - u(\tau y)| < \varepsilon$ and this shows that the function u is uniformly continuous (with respect to the metric d) on the set

$$W = \bigcup_{\tau \in F} B \cap (\{\tau\} \times I^\mathbb{N}).$$

Since $\overline{W} = S^\mathbb{Z}$ (as $m(W) = 1$) and since u is independent of the past, we conclude that u extends continuously to $S^\mathbb{Z}$. Since both sides of (8.3) are continuous functions, and the equality in (8.3) holds on the dense set $W \cap \sigma^{-1}(W)$, we are done.   □

8.2 Classical Conformal Expanding Random Systems

If a classical system is conformal in the sense of Definition 5.1 and if the potential is of the form $\varphi = -t \log|f'|$ for some $t \in \mathbb{R}$ then we will call it *classical conformal expanding random system*

Theorem 8.6 *Suppose* $f : \mathscr{J} \to \mathscr{J}$ *is a classical conformal expanding random system. Then the following hold.*

(a) *The asymptotic variance* $\sigma^2(P(h))$ *exists.*
(b) *If* $\sigma^2(P(h)) > 0$, *then the system* $f : \mathscr{J} \to \mathscr{J}$ *is essential,* $\mathscr{H}^h(\mathscr{J}_x) = 0$ *and* $\mathscr{P}^h(\mathscr{J}_x) = +\infty$ *for m-a.e.* $x \in I^\mathbb{Z}$.
(c) *If, on the other hand,* $\sigma^2(P(h)) = 0$, *then the system* $f : \mathscr{J} \to \mathscr{J}$, *reduced in the base to the topological support of* m *(equal to* $\mathrm{supp}(m_0)^\mathbb{Z}$), *is quasi-deterministic, and then for every* $x \in \mathrm{supp}(m)$, *we have:*

   (c1) v_x^h *is a geometric measure with exponent* h.
   (c2) *The measures* v_x^h, $\mathscr{H}^h|_{\mathscr{J}_x}$, *and* $\mathscr{P}^h|_{\mathscr{J}_x}$ *are all mutually equivalent with Radon–Nikodym derivatives separated away from zero and infinity independently of* $x \in I^\mathbb{Z}$ *and* $y \in \mathscr{J}_x$.
   (c3) $0 < \mathscr{H}^h(\mathscr{J}_x), \mathscr{P}^h(\mathscr{J}_x) < +\infty$ *and* $\mathrm{HD}(\mathscr{J}_x) = h$.

Proof. It follows from Corollary 8.4 that the asymptotic variance $\sigma^2(P(h))$ exists. Combining this corollary (the Law of Iterated Logarithm) with Remark 5.5, we conclude that the system $f : \mathscr{J} \to \mathscr{J}$ is essential. Hence, item (b) follows from Theorem 5.7(a). If, on the other hand, $\sigma^2(P(h)) = 0$, then the system $f : \mathscr{J} \to \mathscr{J}$, reduced in the base to the topological support of m (equal to supp$(m_0)^{\mathbb{Z}}$), is quasi-deterministic because of Proposition 8.5, Theorem 8.3 ($P(h) \in \mathscr{H}_\kappa$), and Remark 5.6. Items (c1)–(c4) follow now from Theorem 5.7(b1)–(b4). We are done. $\qquad\square$

As a consequence of this theorem we get the following.

Theorem 8.7 *Suppose $f : \mathscr{J} \to \mathscr{J}$ is a classical conformal expanding random system. Then the following hold:*

(a) *Suppose that for every $x \in I^{\mathbb{Z}}$, the fiber $\mathscr{J}_x$ is connected. If there exists at least one $w \in$ supp(m) such that HD$(\mathscr{J}_w) > 1$, then HD$(\mathscr{J}_x) > 1$ for m-a.e. $x \in I^{\mathbb{Z}}$.*

(b) *Let d be the dimension of the ambient Riemannian space Y. If there exists at least one $w \in$ supp(m) such that HD$(\mathscr{J}_w) < d$, then HD$(\mathscr{J}_x) < d$ for m-a.e. $x \in I^{\mathbb{Z}}$.*

Proof. Let us proof first item (a). By Theorem 8.6(a) the asymptotic variance $\sigma^2(P(h))$ exists. If $\sigma^2(P(h)) > 0$, then by Theorem 8.6(a) the system $f : \mathscr{J} \to \mathscr{J}$ is essential. Thus the proof is concluded in exactly the same way as the proof of Theorem 5.8(3). If, on the other hand, $\sigma^2(P(h)) = 0$, then the assertion of (a) follows from Theorem 8.6(c4) and the fact that HD$(\mathscr{J}_w) > 1$ and $w \in$ supp(m).

Let us now prove item (b). If $\sigma^2(P(h)) > 0$, then, as in the proof of item (a), the claim is proved in exactly the same way as the proof of Theorem 5.8(4). If, on the other hand, $\sigma^2(P(h)) = 0$, then the assertion of (b) follows from Theorem 8.6(c4) and the fact that HD$(\mathscr{J}_w) < d$ and $w \in$ supp(m). We are done. $\qquad\square$

8.3 Complex Dynamics and Brück and Büger Polynomial Systems

We now want to describe some classes of examples coming from complex dynamics. They will be classical conformal expanding random systems as well as G-systems defined later in this section. Indeed, having a sequence of rational functions $F = \{f_n\}_{n=0}^{\infty}$ on the Riemann sphere $\hat{\mathbb{C}}$ we say that a point $z \in \hat{\mathbb{C}}$ is a member of the Fatou set of this sequence if and only if there exists an open set U_z containing z such that the family of maps $\{f_n|_{U_z}\}_{n=0}^{\infty}$ is normal in the sense of Montel. The Julia set $\mathscr{J}(F)$ is defined to be the complement (in $\hat{\mathbb{C}}$) of the Fatou set of F. For every $k \geq 0$ put $F_k = \{f_{k+n}\}_{n=0}^{\infty}$ and observe that

$$\mathscr{J}(F_{k+1}) = f_k(\mathscr{J}(F_k)). \tag{8.4}$$

Now, consider the maps

$$f_c(z) = f_{d,c}(z) = z^d + c, \quad d \geq 2.$$

Notice that for every $\varepsilon > 0$ there exists $\delta_\varepsilon > 0$ such that if $|c| \leq \delta_\varepsilon$, then

$$f_c(\overline{B}(0, \varepsilon)) \subset \overline{B}(0, \varepsilon).$$

Consequently, if $\omega \in \overline{B}(0, \varepsilon)^{\mathbb{Z}}$, then $\mathcal{J}(\{f_{\omega_n}\}_{n=0}^\infty) \subset \{z \in \mathbb{C} : |z| \geq \varepsilon\}$ and

$$|f'_{\omega_k}(z)| \geq d\varepsilon^{d-1} \tag{8.5}$$

for all $z \in \mathcal{J}(\{f_{\omega_{k+n}}\}_{n=0}^\infty)$. Let $\delta(d) = \sup\left\{\delta_\varepsilon : \varepsilon > \sqrt[d-1]{1/d}\right\}$. Fix $0 < \delta < \delta(d)$. Then there exists $\varepsilon > \sqrt[d-1]{1/d}$ such that $\delta < \delta_\varepsilon$. Therefore, by (8.5),

$$|f'_{\omega_k}(z)| \geq d\varepsilon^{d-1} \tag{8.6}$$

for all $\omega \in \overline{B}(0, \delta)^{\mathbb{Z}}$, all $k \geq 0$ and all $z \in \mathcal{J}(\{f_{\omega_{k+n}}\}_{n=0}^\infty)$. A straight calculation ([8], p. 349) shows that $\delta(2) = 1/4$. Keep $0 < \delta < \delta(d)$ fixed. Let

$$\mathcal{F}_{d,\delta} = \{f_{d,c} : c \in \overline{B}(0, \delta)\}.$$

Consider an arbitrary ergodic measure-preserving transformation $\theta : X \to X$. Let m be the corresponding invariant probability measure. Let also $H : X \to \mathcal{F}_{d,\delta}$ be an arbitrary measurable function. Set $f_{d,x} = H(x)$ for all $x \in X$. For every $x \in X$ let $\mathcal{J}_x$ be the Julia set of the sequence $\{f_{\theta^n(x)}\}_{n=0}^\infty$, and then $\mathcal{J} = \bigcup_{x \in X} \mathcal{J}_x$. Note that, because of (8.4), $f_{d,x}(\mathcal{J}_x) = \mathcal{J}_{\theta(x)}$. Thus, the map

$$f_{d,\delta,\theta,H}(x, y) = (\theta(x), f_{d,x}(y)), \quad x \in X, \ y \in \mathcal{J}_x, \tag{8.7}$$

defines a skew product map in the sense of Chap. 2.2 of our paper. In view of (8.7), when $\theta : X \to X$ is invertible, $f_{d,\delta,\theta,H}$ is a distance expanding random system, and, since all the maps f_x are conformal, $f_{d,\delta,\theta,H}$ is a conformal measurably expanding system in the sense of Definition 5.1. As an immediate consequence of Theorem 5.2 we get the following.

Theorem 8.8 *Let $\theta : X \to X$ be an invertible measurable map preserving a probability measure m. Fix an integer $d \geq 1$ and $0 < \delta < \delta(d)$. Let $H : X \to \mathcal{F}_{d,\delta}$ be an arbitrary measurable function. Finally, let $f_{d,\delta,\theta,H}$ be the distance expanding random system defined by formula (8.7). Then for almost all $x \in X$ the Hausdorff dimension of the Julia set $\mathcal{J}_x$ is equal to the unique zero of the expected value of the pressure function.*

Theorem 8.9 *For the conformal measurably expanding systems $f_{d,\delta,\theta,H}$ defined in Theorem 8.8 the multifractal theorem, Theorem 6.4 holds.*

We now define and deal with Brück and Büger polynomial systems. We still keep $d \geq 2$ and $0 < \delta < \delta(d)$ fixed. Let $X = B(0, \delta)^{\mathbb{Z}}$ and let

$$\theta : B(0, \delta)^{\mathbb{Z}} \to B(0, \delta)^{\mathbb{Z}}$$

to be the shift map denoted in the sequel by σ. Consider any Borel probability measure m_0 on $\overline{B}(0, \delta)$ which is different from δ_0, the Dirac δ measure supported at 0. Define $H : X \to \mathscr{F}_{d,\delta}$ by the formula $H(\omega) = f_{d,\omega_0}$. The corresponding skew-product map $f_{d,\delta} : \mathscr{J} \to \mathscr{J}$ is then given by the formula:

$$f_{d,\delta}(\omega, z) = (\sigma(\omega), f_{d,\omega_0}(z)) = (\sigma(\omega), z^d + \omega_0),$$

and $f_{d,\delta,\omega}(z) = z^d + \omega_0$ acts from $\mathscr{J}_\omega$ to $\mathscr{J}_{\sigma(\omega)}$, where $\mathscr{J}_\omega = \mathscr{J}((f_{d,\omega_n})_{n=0}^\infty)$. Then $f : \mathscr{J} \to \mathscr{J}$ is called *Brück and Büger polynomial systems.* Clearly, $f : \mathscr{J} \to \mathscr{J}$ is a classical conformal expanding random system.

In [8] Brück speculated on p. 365 that if $\delta < 1/4$ and m_0 is the normalized Lebesgue measure on $\overline{B}(0, \delta)$, then $\mathrm{HD}(\mathscr{J}_\omega) > 1$ for m_+-a.e. $\omega \in \overline{B}(0, \delta)^{\mathbb{N}}$ with respect to the skew-product map

$$(\omega, z) \mapsto (\sigma(\omega), z^2 + \omega_0).$$

In [9] this problem was explicitly formulated by Brück and Büger as Question 5.4. Below (Theorem 8.10) we prove a more general result (with regard the measure on $\overline{B}(0, \delta)$ and the integer $d \geq 2$ being arbitrary), which contains the positive answer to the Brück and Büger question as a special case. In [8] Brück also proved that if $\delta < 1/4$ and the above skew product is considered then $\lambda_2(\mathscr{J}_\omega) = 0$ for all $\omega \in \overline{B}(0, \delta)^{\mathbb{N}}$, where λ_2 denotes the planar Lebesgue measure on $\mathbb{C}$. As a special case of Theorem 8.10 below we get a partial strengthening of Brück's result saying that $\mathrm{HD}(\mathscr{J}_\omega) < 2$ for m_+-a.e. $\omega \in \overline{B}(0, \delta)^{\mathbb{N}}$. Our results are formulated for the product measure m on $\overline{B}(0, \delta)^{\mathbb{Z}}$, but as m_+ is the projection from $\overline{B}(0, \delta)^{\mathbb{Z}}$ to $\overline{B}(0, \delta)^{\mathbb{N}}$ and as the Julia sets $\mathscr{J}_\omega$, $\omega \in \overline{B}(0, \delta)^{\mathbb{Z}}$ depend only on $\omega|_0^{+\infty}$, i.e. on the future of ω, the analogous results for m_+ and $\overline{B}(0, \delta)^{\mathbb{N}}$ follow immediately. Proving what we have just announced, note that if $\omega_0 \in \mathrm{supp}(m_0) \setminus \{0\}$, then

$$\mathrm{HD}(\mathscr{J}_{\omega_0^\infty})) = \mathrm{HD}(\mathscr{J}(f_{\omega_0})) \in (1, 2)$$

(the equality holds already on the level of sets: $\mathscr{J}_{\omega_0^\infty} = \mathscr{J}(f_{\omega_0})$), and by [9], all the sets $\mathscr{J}_\omega$, $\omega \in \overline{B}(0, \delta)^{\mathbb{Z}}$, are Jordan curves. Hence, since $f : \mathscr{J} \to \mathscr{J}$ is a classical conformal expanding random system, as an immediate application of Theorem 8.7 we get the following.

Theorem 8.10 *If $d \geq 2$ is an integer, $0 < \delta < \delta(d)$, the skew-product map $f_{d,\delta} :$ $\mathscr{J} \to \mathscr{J}$ is given by the formula*

$$f_{d,\delta}(\omega, z) = (\sigma(\omega), f_{d,\omega_0}(z)) = (\sigma(\omega), z^d + \omega_0),$$

and m_0 is an arbitrary Borel probability measure on $\overline{B}(0, \delta)$, different from δ_0, the Dirac δ measure supported at 0, then for m-almost every $\omega \in \overline{B}(0, \delta)^{\mathbb{Z}}$ we have $1 < \mathrm{HD}(\mathcal{J}_\omega) < 2$.

8.4 Denker–Gordin Systems

We now want to discuss another class of expanding random maps. This is the setting from [12]. In order to describe this setting suppose that X_0 and Z_0 are compact metric spaces and that $\theta_0 : X_0 \to X_0$ and $T_0 : Z_0 \to Z_0$ are open topologically exact distance expanding maps in the sense as in [24]. We assume that T_0 is a skew-product over Z_0, i.e. for every $x \in X_0$ there exists a compact metric space $\mathcal{J}_x$ such that $Z_0 = \bigcup_{X \in X_0} \{x\} \times \mathcal{J}_x$ and the following diagram commutes:

$$
\begin{array}{ccc}
Z_0 & \xrightarrow{\;\;T_0\;\;} & Z_0 \\
\Big\downarrow{\scriptstyle \pi} & & \Big\downarrow{\scriptstyle \pi} \\
X_0 & \xrightarrow{\;\;\theta_0\;\;} & X_0
\end{array}
$$

where $\pi(x, y) = x$ and the projection $\pi : Z_0 \to X_0$ is an open map. Additionally, we assume that there exists L such that

$$d_{X_0}(\theta_0(x), \theta_0(x')) \le L d_X(x, x') \tag{8.8}$$

for all $x \in X$ and that there exists $\xi_1 > 0$ such that, for all x, x' satisfying $d_{X_0}(x, x') < \xi_1$ there exist y, y' such that

$$d\big((x, y), (x', y')\big) < \xi. \tag{8.9}$$

We then refer to $T_0 : Z_0 \to Z_0$ and $\theta_0 : X_0 \to X_0$ as a DG-system. Note that

$$T_0(\{x\} \times \mathcal{J}_x) \subset \{\theta_0(x)\} \times \mathcal{J}_{\theta_0(x)}$$

and this gives rise to the map $T_x : \mathcal{J}_x \to \mathcal{J}_{\theta_0(x)}$.

Since T_0 is distance expanding, conditions uniform openness, measurably expanding measurability of the degree, topological exactness (see Chap. 2) hold with some constants $\gamma_x \ge \gamma > 1$, $\deg(T_x) \le N_1 < +\infty$ and the number $n_r = n_r(x)$ in fact independent of x. Scrutinizing the proof of Remark 2.9 in [12] one sees that Lipschitz continuity (Denker and Gordin assume differentiability) suffices for it to go through and Lipschitz continuity is incorporated in the definition of expanding maps in [24]. Now assume that $\phi : Z \to \mathbb{R}$ is a Hölder continuous map. Then the hypothesis of Theorems 2.10, 3.1, and 3.2 from [12] are satisfied. Their claims are summarized in the following.

Theorem 8.11 *Suppose that* $T_0 : Z_0 \to Z_0$ *and* $\theta_0 : X_0 \to X_0$ *form a DG-system and that* $\phi : Z \to \mathbb{R}$ *is a Hölder continuous potential. Then there exists a Hölder continuous function* $P(\phi) : X_0 \to \mathbb{R}$, *a measurable collection* $\{v_x\}_{x \in X_0}$ *and a continuous function* $q : Z_0 \to [0, +\infty)$ *such that*

(a) $v_{\theta_0(x)}(A) = \exp(P_x(\phi)) \int_A e^{-\phi_x} dv_x$ *for all* $x \in X_0$ *and all Borel sets* $A \subset \mathscr{J}_x$ *such that* $T_x|_A$ *is one-to-one.*
(b) $\int_{\mathscr{J}_x} q_x dv_x = 1$ *for all* $x \in X_0$.
(c) *Denoting for every* $x \in X_0$ *by* μ_x *the measure* $q_x v_x$ *we have*

$$\sum_{w \in \theta_0^{-1}(x)} \mu_w(T_w^{-1}(A)) = \mu_x(A) \qquad \text{for every Borel set} \quad A \subset \mathscr{J}_x .$$

This would mean that we got all the objects produced in Chap. 3 of our paper. However, the map $\theta_0 : X_0 \to X_0$ need not be, and apart from the case when X_0 is finite, is not invertible. But to remedy this situation is easy. We consider the projective limit (Rokhlin's natural extension) $\theta : X \to X$ of $\theta_0 : X_0 \to X_0$. Precisely,

$$X = \{(x_n)_{n \leq 0} : \theta_0(x_n) = x_{n+1} \; \forall n \leq -1\}$$

and

$$\theta\big((x_n)_{n \leq 0}\big) = (\theta_0(x_n))_{n \leq 0}.$$

Then $\theta : X \to X$ becomes invertible and the diagram

$$
\begin{array}{ccc}
X & \xrightarrow{\;\;\theta\;\;} & X \\
\downarrow{\scriptstyle p} & & \downarrow{\scriptstyle p} \\
X_0 & \xrightarrow{\;\;\theta_0\;\;} & X_0
\end{array}
\qquad (8.10)
$$

commutes, where $p((x_n)_n \leq 0) = x_0$. If in addition, as we assume from now on, the space X is endowed with a Borel probability θ_0-invariant ergodic measure m_0, then there exists a unique θ-invariant probability measure m such that $m \circ \pi^{-1} = m_0$. Let

$$Z := \bigcup_{x \in X} \{x\} \times \mathscr{J}_{x_0}.$$

We define the map $T : Z \to Z$ by the formula $T(x, y) = (\theta(x), T_{x_0}(y))$ and the potential $X \ni x \mapsto \phi(x_0)$ from X to $\mathbb{R}$. We keep for it the same symbol ϕ. Clearly the quadruple (T, θ, m, ϕ) is a Hölder fiber system as defined in Chap. 2 of our paper. It follows from Theorem 8.11 along with the definition of θ a commutativity of the diagram (8.10) for $x \in X$ all the objects $P_x(\phi) = P_{x_0}(\phi)$, $\lambda_x = \exp(P_x(\phi))$, $q_x = q_{x_0}$, $v_x = v_{x_0}$, and $\mu_x = \mu_{x_0}$ enjoy all the properties required in

Theorems 3.1 and 3.2; in particular they are unique. From now on we assume that the measure m is a Gibbs state of a Hölder continuous potential on X (having nothing to do with ϕ or $P(\phi)$; it is only needed for the Law of Iterated Logarithm to hold). We call the quadruple (T, θ, m, ϕ) DG*-system.

The following Hölder continuity theorem appeared in the paper [12]. We provide here an alternative proof under weaker assumptions.

Theorem 8.12 If $d_X(x, x') < \xi$, then $|\lambda_x - \lambda_{x'}| \le H d_X^\alpha(x, x')$.

Proof. Let n be such that

$$d_X(\theta^{2n-1}(x), \theta^{2n-1}(x')) < \xi_1 \quad \text{and} \quad d_X(\theta^{2n}(x), \theta^{2n}(x')) \ge \xi_1. \quad (8.11)$$

Let $z \in T^{-2n+1}(y)$ and $z' \in T^{-2n+1}(y')$. Then for all $k = 0, \ldots, n-1$

$$|\varphi(T^k(z)) - \varphi(T^k(z'))| \le C d^\alpha(T^k(z), T^k(z')) \le C \gamma^{-\alpha n} \gamma^{-\alpha(n-k-1)} \xi.$$

Then

$$|S_n \varphi(z) - S_n \varphi(z')| \le \frac{C \xi \gamma^{-\alpha n}}{1 - \gamma^{-\alpha}}.$$

Put $C' := C \xi / (1 - \gamma^{-\alpha})$. Then

$$\left| \log \frac{\mathscr{L}_x^n \mathbb{1}(w)}{\mathscr{L}_{x'}^n \mathbb{1}(w')} \right| \le C' \gamma^{-\alpha n} \quad \text{and} \quad \left| \log \frac{\mathscr{L}_{\theta(x)}^{n-1} \mathbb{1}(w)}{\mathscr{L}_{\theta(x')}^{n-1} \mathbb{1}(w')} \right| \le C' \gamma^{-\alpha n}.$$

Then

$$\left| \log \frac{\mathscr{L}_x^n \mathbb{1}(w)}{\mathscr{L}_{\theta(x)}^{n-1} \mathbb{1}(w)} - \log \frac{\mathscr{L}_{x'}^n \mathbb{1}(w')}{\mathscr{L}_{\theta(x')}^{n-1} \mathbb{1}(w')} \right| \le 2C' \gamma^{-\alpha n}. \quad (8.12)$$

Let $\alpha' := (\alpha \log \gamma)/(2 \log L)$. Then by (8.11)

$$\gamma^{-n\alpha} = L^{-2n\alpha'} \le \frac{(d(\theta^{2n}(x), \theta^{2n}(x')))^{\alpha'}}{\xi_1^{\alpha'} L^{-2n\alpha'}} \le \frac{(d(x, x'))^{\alpha'}}{\xi_1^{\alpha'}}.$$

Then (8.12) finishes the proof. $\qquad \square$

Since the map $\theta_0 : X_0 \to X_0$ is expanding, since m is a Gibbs state, and since $P(\phi) : X_0 \to \mathbb{R}$ is Hölder continuous, it is well known (see [24] for example) that the following asymptotic variance exists:

$$\sigma^2(P(\phi)) = \lim_{n \to \infty} \frac{1}{n} \int \left(S_n(P(\phi)) - n \mathscr{E} P(\phi) \right)^2 dm.$$

The following theorem of Livsic flavor is (by now) well known (see [24]).

Theorem 8.13 *Suppose* (T, θ, m, ϕ) *is a DG*-system. Then the following are equivalent.*

(a) $\sigma^2(P(\phi)) = 0$.
(b) *The function* $P(\phi)$ *is cohomologous to a constant in the class of real-valued continuous functions on* X *(resp.* X_0*), meaning that there exists a continuous function* $u : X \to \mathbb{R}$ *(resp.* $u : X_0 \to \mathbb{R}$*) such that*

$$P(\phi) - (u - u \circ \theta) \quad (resp. \ P(\phi) - (u - u \circ \theta_0))$$

is a constant.
(c) *The function* $P(\phi)$ *is cohomologous to a constant in the class of real-valued Hölder continuous functions on* X *(resp.* X_0*), meaning that there exists a Hölder continuous function* $u : X \to \mathbb{R}$ *(resp.* $u : X \to \mathbb{R}$*) such that*

$$P(\phi) - (u - u \circ \theta) \quad (resp. \ P(\phi) - (u - u \circ \theta_0))$$

is a constant.
(d) *There exists* $R \in \mathbb{R}$ *such that* $P_x^n(\phi) = nR$ *for all* $n \geq 1$ *and all periodic points* $x \in X$ *(resp.* X_0*).*

As a matter of fact such theorem is formulated in [24] for non-invertible (θ_0) maps only but it also holds for the Rokhlin's natural extension θ. The following theorem follows directly from [24] and Theorem 8.11 (Hölder continuity of $P(\phi)$).

Theorem 8.14 *(The Law of Iterated Logarithm) If* (T, θ, m, ϕ) *is a DG*-system and if* $\sigma^2(P(\phi)) > 0$, *then m-a.e. we have*

$$-\sqrt{2\sigma^2(P(\phi))} = \liminf_{n \to \infty} \frac{P_x^n(\phi) - n\mathscr{E}P(\phi)}{\sqrt{n \log \log n}} \leq \limsup_{n \to \infty} \frac{P_x^n(\phi) - n\mathscr{E}P(\phi)}{\sqrt{n \log \log n}} = \sqrt{2\sigma^2(P(\phi))}.$$

8.5 Conformal DG*-Systems

Now we turn to geometry. This section dealing with, below defined, conformal DG*-systems is a continuation of the previous one in the setting of conformal systems. We shall show that these systems naturally split into essential and quasi-deterministic, and will establish their fractal and geometric properties. Suppose that (f_0, θ_0) is a DG-system endowed with a Gibbs measure m_0 at the base. Suppose also that this system is a random conformal expanding repeller in the sense of Chap. 5 and that the function $\phi : Z \to \mathbb{R}$ given by the formula

$$\phi(x, y) = -\log |f_x'(y)|,$$

is Hölder continuous.

Definition 8.15 *The corresponding system* $(f, \theta, m) = (f, \theta, m, \phi)$ *(with θ the Rokhlin natural extension of θ_0 as described above) is called conformal DG*-system.*

For every $t \in \mathbb{R}$ the potential $\phi_t = t\phi$, considered in Chap. 5, is also Hölder continuous. As in Chap. 5 denote its topological pressure by $P(t)$. Recall that h is a unique solution to the equation $\mathscr{E}P(t) = 0$. By Theorem 5.2 (Bowen's Formula) $\mathrm{HD}(\mathscr{J}_x) = h$ for m-a.e. $x \in X$. As an immediate consequence of Theorems 5.7, 8.14, and Remark 5.6, we get the following.

Theorem 8.16 *Suppose* $(f, \theta, m) = (f, \theta, m, \phi)$ *is a random conformal DG*-system.*

(a) *If $\sigma^2(P(h)) > 0$, then the system (f, θ, m) is essential, and then*

$$\mathscr{H}^h(\mathscr{J}_x) = 0 \quad \text{and} \quad \mathscr{P}^h(\mathscr{J}_x) = +\infty.$$

(b) *If, on the other hand, $\sigma^2(P(h)) = 0$, then $(f, \theta, m) = (f, \theta, m, \phi)$ is quasi-deterministic, and then for every $x \in X$, we have that v_x^h is a geometric measure with exponent h and, consequently, the geometric properties (GM1)–(GM3) hold.*

Exactly as Corollary 5.8 is a consequence of Theorem 5.7, the following corollary is a consequence of Theorem 8.16.

Corollary 8.17 *Suppose* $(f, \theta, m) = (f, \theta, m, \phi)$ *is a conformal DG*-system and $\sigma^2(P(h)) > 0$. Then the system (f, θ, m) is essential, and for m-a.e. $x \in X$ the following hold.*

1. *The fiber $\mathscr{J}_x$ is not bi-Lipschitz equivalent to any deterministic nor quasi-deterministic self-conformal set.*
2. *$\mathscr{J}_x$ is not a geometric circle nor even a piecewise smooth curve.*
3. *If $\mathscr{J}_x$ has a non-degenerate connected component (for example if $\mathscr{J}_x$ is connected), then*
$$h = \mathrm{HD}(\mathscr{J}_x) > 1.$$
4. *Let d be the dimension of the ambient Riemannian space Y. Then $\mathrm{HD}(\mathscr{J}_x) < d$.*

Now, in the same way as Theorem 8.7 is a consequence of Theorem 8.6, Corollary 8.17 yields the following.

Theorem 8.18 *Suppose* $(f, \theta, m) = (f, \theta, m, \phi)$ *is a conformal DG*-system. Then the following hold.*

(a) *Suppose that for every $x \in X$, the fiber $\mathscr{J}_x$ is connected. If there exists at least one $w \in \mathrm{supp}(m)$ such that $\mathrm{HD}(\mathscr{J}_w) > 1$, then*

$$\mathrm{HD}(\mathscr{J}_x) > 1 \quad \text{for } m\text{-a.e.} \quad x \in I^{\mathbb{Z}}.$$

(b) *Let d be the dimension of the ambient Riemannian space Y. If there exists at least one $w \in X$ such that $\mathrm{HD}(\mathscr{J}_w) < d$, then $\mathrm{HD}(\mathscr{J}_x) < d$ for m-a.e. $x \in X$.*

We end this subsection and the entire section with a concrete example of a conformal DG*-system. In particular, the three above results apply to it. Let

$$X := S^1_{\delta_d} = \{z \in \mathbb{C} : |z| = \delta\}.$$

Fix an integer $k \geq 2$. Define the map $\theta_0 : X \to X$ by the formula $\theta_0(x) = \delta^{1-k}x^k$. Then $\theta_0'(x) = k\delta^{1-k}x^{k-1}$ and therefore $|\theta_0'(x)| = k \geq 2$ for all $x \in X$. The normalized Lebesgue measure λ_0 on X is invariant under θ_0. Define the map $H :$ $X \to \mathscr{F}_d$ by setting $H(x) = f_x$. Then

$$f_{\theta_0,H,0}(x, y) = (k\delta^{1-k}x^{k-1}, g^d + x).$$

Note that $\left(f_{\theta_0,H,0}, \theta_0, \lambda_0\right)$ is a uniformly conformal DG-system and let $(f_{\theta,H}, \theta, \lambda)$ be the corresponding random conformal G-system, both in the sense of Chap. 5. Theorems 8.16, 8.18, and Corollary 8.17 apply.

8.6 Random Expanding Maps on Smooth Manifold

We now complete the previous examples with some remarks on random maps on smooth manifolds. Let (M, ρ) be a smooth compact Riemannian manifold. We recall that a differentiable endomorphism $f : M \to M$ is expanding if there exists $\gamma > 1$ such that

$$\|f_x'(v)\| \geq \gamma \|v\| \qquad \text{for all } x \in M \text{ and all } v \in T_x M .$$

The largest constant $\gamma > 1$ enjoying this property is denoted by $\gamma(f)$. If $\gamma > 1$, we denote by $\mathscr{E}_\gamma(M)$ the set of all expanding endomorphisms of M for which $\gamma(f) \geq \gamma$. We also set

$$\mathscr{E}(M) = \bigcup_{\gamma > 1} \mathscr{E}_\gamma(M),$$

i.e. $\mathscr{E}(M)$ is the set of all expanding endomorphisms of M.

8.7 Topological Exactness

We shall prove the following.

Proposition 8.19 *Suppose that M is a connected and compact manifold and that $f_n \in \mathscr{E}(M), n \geq 1$, are endomorphisms such that*

$$lim_{n\to\infty} \prod_{j=1}^n \gamma(f_j) = +\infty .$$

Denote $F_k = f_k \circ f_{k-1} \circ \ldots \circ f_1$, $k \geq 1$. Then, for every $r > 0$ there exist $k \geq 1$ such that

$$F_k(B(x, r)) = M \quad \text{for every } x \in M .$$

In particular, if U is a non-empty open subset of M, then there exists $k \geq 1$ such that $F_k(U) = M$.

Proof. Let $f \in \mathscr{E}(M)$, set $\gamma = \gamma(f)$ and notice first of all that for such a map the implicit function theorem applies and yields that f is an open map. The manifold M being connected, it follows that f is surjective. Moreover, if β is any path starting at a point $y = f(x)$, then there is a lift α starting at x. The expanding property implies that

$$length(\beta) = length(f \circ \alpha) \geq \gamma \, length(\alpha) .$$

In particular, if β is a geodesic between $y = f(x)$ and a point $y' \in M$, then there is a point $x' \in M$ such that $f(x') = y'$ and

$$\rho(y, y') \geq \gamma \, \rho(x, x') .$$

This shows that for every $r > 0$ and every $x \in M$ we have

$$f(B(x, r)) \supset B(f(x), \gamma r) .$$

The proposition follows now from the compactness of M.   □

8.8 Stationary Measures

Let M be an n-dimensional compact Riemannian manifold and let I be a set equipped with a probabilistic measure m_0. To every $a \in I$ we associate a differentiable expanding transformation f_a of M into itself. Put $X = I^{\mathbb{Z}}$ and let m be the product measure induced by m_0. For $x = \ldots a_{-1} a_0 a_1 \ldots$ consider $\varphi_x := -\log |\det f'_{a_0}|$. We assume that all our assumptions are satisfied. Then the measure $\nu = vol_M$ (where vol_M is the normalized Riemannian volume on M) is the fixed point of the operator $\mathscr{L}^*_{x,\varphi}$ with $\lambda_x = 1$. Let q_x be the function given by Theorem 3.1, and let μ_x be the measure determined by $d\mu_x / d\nu_x = q_x$.

We write $I^{\mathbb{Z}} = I^{-\mathbb{N}} \times I^{\mathbb{N}}$ where points from $I^{-\mathbb{N}}$ are denoted by $x^- = \ldots a_{-2} a_{-1}$ and from $I^{\mathbb{N}}$ by $x^+ = a_0 a_1 \ldots$. Then $x^- x^+$ means $x = \ldots a_{-1} a_0 a_1 \ldots$. Note that q_x does not depend on x^+, since nor does $\mathscr{L}^n_{x-n} \mathbb{1}(y)$. Then we can write $q_{x^-} := q_x$ and $\mu_{x^-} := \mu_x$. Since $\mu_x(g \circ f_{a_0}) = \mu_{\theta(x)}$ we have that

$$\mu_{x^-}(g \circ f_a) = \mu_{x^- a}(g) \tag{8.13}$$

for every $a \in I$.

Define a measure μ^* by $d\mu^* = d\mu_x - dm^-(x^-)$ where m^- is the product measure on $I^{-\mathbb{N}}$. Then by (8.13)

$$\int \mu^*(g \circ f_a) dm_0(a) = \int \mu_{x^-}(g \circ f_a) dm^-(x^-)$$

$$= \int \int \mu_{x^-a}(g) dm^-(x^-) dm_0(a) = \mu^*(g).$$

Therefore, μ^* is a stationary measure (see for example [28]).

Chapter 9
Real Analyticity of Pressure

Here we provide, in particular, the real analyticity results that where used in the proof of the real analyticity of the multifractal spectrum (Chap. 6.3). We putted this part at the end of the manuscript since, as already mentioned, it is of different nature. It is heavily based on ideas of Rugh [26] and uses the Hilbert metric on appropriately chosen cones.

9.1   The Pressure as a Function of a Parameter

Here, we will have a careful close look at the measurable bounds obtained in Chap. 3 from which we deduce that the theorems from that section can be proved to hold for every parameter and almost every x (common for all parameters).

In this section we only assume that $T : \mathcal{J} \to \mathcal{J}$ is a measurable expanding random map. Let $\varphi^{(1)}, \varphi^{(2)} \in \mathcal{H}_m(\mathcal{J})$ and let $t = (t_1, t_2) \in \mathbb{R}^2$. Put

$$|t| := \max\{|t_1|, |t_2|\} \quad \text{and} \quad t^* := \max\{1, |t|\}. \tag{9.1}$$

Set $\varphi_t =: t_1 \varphi^{(1)} + t_2 \varphi^{(2)}$ and

$$\varphi := |\varphi^{(1)}| + |\varphi^{(2)}|. \tag{9.2}$$

Fix $\alpha > 0$ and a measurable log-integrable function $H : X \to [0, +\infty)$ such that $\varphi^{(1)}, \varphi^{(2)} \in \mathcal{H}_m^\alpha(\mathcal{J}, H)$. Then for all $x \in X$ and all $y_1, y_2 \in \mathcal{J}_x$, we have

$$|\varphi_{t,x}(y_2) - \varphi_{t,x}(y_1)| \le H_x |t_1| \rho_x^\alpha(y_2, y_1) + H_x |t_2| \rho_x^\alpha(y_2, y_1) \le 2|t| H_x \rho_x^\alpha(y_2, y_1).$$

Therefore $\varphi_t \in \mathcal{H}_m^\alpha(\mathcal{J}, 2|t|H) \subset \mathcal{H}_m^\alpha(\mathcal{J}, 2t^*H)$. Also, for all $x \in X$ and all $y \in \mathcal{J}_x$, we have

V. Mayer et al., *Distance Expanding Random Mappings, Thermodynamical Formalism, Gibbs Measures and Fractal Geometry*, Lecture Notes in Mathematics 2036, DOI 10.1007/978-3-642-23650-1_9, © Springer-Verlag Berlin Heidelberg 2011

$$|S_n\varphi_{t,x}(y)| \leq |t_1||S_n\varphi_x^{(1)}(y)| + |t_2||S_n\varphi_x^{(2)}(y)| \leq |t|||S_n\varphi_x(y)| \leq |t|||S_n\varphi_x||_\infty.$$

This implies

$$||S_n\varphi_{t,x}||_\infty \leq |t|||S_n\varphi_x||_\infty \leq t^*||S_n\varphi_x||_\infty. \tag{9.3}$$

Concerning the potential φ, we get

$$|\varphi_x(y_2) - \varphi_x(y_1)| \leq \left||\varphi_x^{(1)}(y_2) - \varphi_x^{(1)}(y_1)\right| + \left|\varphi_x^{(2)}(y_2) - \varphi_x^{(2)}(y_1)\right| \leq 2H_x\rho_x^\alpha(y_2, y_1).$$

Thus

$$\varphi \in \mathscr{H}_m^\alpha(\mathscr{J}, 2H). \tag{9.4}$$

Denote by C_t, $C_{t,\max}$, $C_{t,\min}$, $D_{\xi,t}$ and $\beta_t(s)$, the respective functions associated to the potential φ_t as in Chap. 3.2. If the index t is missing, these numbers, as usually, refer to the potential φ given by (9.2). Using (9.3) and (9.4), we then immediately get

$$D_{\xi,t}(x) \geq D_{\xi,\varphi}^{t^*}, \tag{9.5}$$

$$C_t(x) \leq \exp(Q_x(2t^*H)) \max_{0 \leq k \leq j} \left\{\exp(2t^*||S_k\varphi_{x-k}||_\infty)\right\}$$

$$\leq \left(\exp(Q_x(2H)) \max_{0 \leq k \leq j} \left\{\exp(2||S_k\varphi_{x-k}||_\infty)\right\}\right)^{t^*} = C_\varphi^{t^*}, \tag{9.6}$$

$$C_{t,\min}(x) \geq \exp(-Q_x(2t^*H)) \exp(-2t^*||S_n\varphi_x||_\infty) = C_{\min}(x)^{t^*}, \tag{9.7}$$

$$C_{t,\max}(x) = \exp(Q_x(2t^*H)) \deg(T_x^n) \exp(2t^*||S_n\varphi_x||_\infty) \leq C_{\max}(x)^{t^*}, \tag{9.8}$$

and therefore,

$$\beta_{t,x}(s) \geq \left(\frac{C_{\min}(x)}{C_\varphi(x)}\right)^{t^*} \frac{(s-1)2t^*H_{x-1}\gamma_{x-1}^{-\alpha}}{4t^*sQ_x} = \left(\frac{C_{\min}(x)}{C_\varphi(x)}\right)^{t^*} \frac{(s-1)H_{x-1}\gamma_{x-1}^{-\alpha}}{2sQ_x}$$

$$\geq \left(\frac{C_{\min}(x)}{C_\varphi(x)}\right)^{t^*} \left(\frac{(s-1)H_{x-1}\gamma_{x-1}^{-\alpha}}{2sQ_x}\right)^{t^*} = \beta_x^{t^*}(s).$$

Finally we are going to look at the function $A(x)$ and the constant B obtained in Proposition 3.17. We fix the set

$$G := \{x : \beta_x \geq M \text{ and } j(x) \leq J\}$$

as defined by (3.35). Note that by (9.1), for $x \in G$ we have, $\beta_{x,t} \geq M^{t^*}$. Denote by G'_- the corresponding visiting set for backward iterates of θ, and by $(n_k)_1^\infty$ the corresponding visiting sequence. In particular $\lim_{k \to \infty} \frac{k}{n_k} \geq \frac{3}{4J}$. Putting $B_t = \sqrt[4J^2]{1 - M^{t^*}}$ and

$$A_t(x) := \max\{2C_{\max}^{t^*}(x)B_t^{-Jk_x^*}, C_\varphi^{t^*}(x) + C_{\max}^{t^*}(x)\},$$

as an immediate consequence of Proposition 3.17 and its proof along with our estimates above, we obtain the following.

Proposition 9.1 *For every $t \in \mathbb{R}^2$, for every $x \in G'_-$, and every $g_x \in \Lambda^s_{t,x}$*

$$\|\tilde{\mathscr{L}}^n_{x-n,t} g_{x-n} - q_{t,x}\|_\infty \leq A_t(x) B^n_t.$$

More generally, if $g_x \in \mathscr{H}^\alpha(\mathscr{J}_x)$, then

$$\left\|\hat{\mathscr{L}}^n_{t,x} g_x - \left(\int g_x d\mu_{t,x}\right)\mathbb{1}\right\|_\infty$$

$$\leq C^{t*}_\varphi(\theta^n(x))\left(\int |g_x| d\mu_{t,x} + 4\frac{v_\alpha(g_x q_{t,x})}{t_* Q_x}\right) A_{t_*}(\theta^n(x)) B^n_{t_*}.$$

In here and in the sequel, by $q_{t,x}$, $\Lambda^s_{t,x}$ and $\mathscr{L}_{t,x}$ we denote the respective objects for the potential φ_t.

Remark 9.2 *It follows from the estimates of all involved measurable functions, that, for $R > 0$ and $t \in \mathbb{R}$ such that $|t| \leq R$, the functions A_t and B_t in Proposition 9.1 can be replaced by $A_{\max\{R,1\}}$ and $B_{\max\{R,1\}}$, respectively.*

Now, let us look at Proposition 3.19. Similarly as with the set G, we consider the set X_A defined by (3.38) with $A(x)$ generated by φ. So, if $x \in X_A$, then $A_t(x) \leq \mathscr{A}_t$ for some finite number $\mathscr{A}_t$ which depends on t. Denote by $X'_{A,+}$ the corresponding visiting set intersected with G'_+. Therefore, the following is a consequence of the proof of Proposition 3.19 and the formula (3.43).

Proposition 9.3 *For every $R > 0$, every $x \in X'_{A,+}$, and every $g_x \in \mathscr{C}(\mathscr{J}_x)$ we have that*

$$\lim_{n\to\infty} \sup_{|t|\leq R} \left\{\left\|\hat{\mathscr{L}}^n_{t,x} g_x - \left(\int g_x d\mu_{t,x}\right)\mathbb{1}_{\theta^n(x)}\right\|_\infty\right\} = 0.$$

Moreover, we obtain the following consequence of Lemma 3.28 and (9.5).

Lemma 9.4 *There exist a set $X' \subset X$ of full measure, and a measurable function $X \ni x \mapsto D_1(x)$ with the following property. Let $x \in X'$, let $w \in \mathscr{J}_x$, and let $n \geq 0$. Put $y = (x, w)$. Then*

$$(D_1(\theta^n(x)))^{-t*} \leq \frac{v_{t,x}(T_y^{-n}(B(T^n(y), \xi)))}{\exp(S_n\varphi_t(y) - S_n P_x(\varphi_t))} \leq (D_1(\theta^n(x)))^{t*}$$

for all $t \in \mathbb{R}^2$.

For all $t \in \mathbb{R}^2$ set

$$\mathscr{E}P(t) := \mathscr{E}P(\varphi_t).$$

We now shall prove the following.

Lemma 9.5 *The function $\mathscr{E}P : \mathbb{R}^2 \to \mathbb{R}$ is convex, and therefore, continuous. There exists a measurable set $X_{\mathscr{E}}'$ such that $m(X_{\mathscr{E}}') = 1$ and for all $x \in X_{\mathscr{E}}'$ and all $t \in \mathbb{R}^2$, the limit*

$$\lim_{n\to\infty} \frac{1}{n} \log \mathscr{L}_{t,x}^n \mathbb{1}(w_n) \tag{9.9}$$

exists, and is equal to $\mathscr{E}P(t)$.

Proof. By Lemmas 4.6 and 3.27 we know that for every $t \in \mathbb{R}^2$ there exists a measurable X_t' with $m(X_t') = 1$ and such that

$$\lim_{n\to\infty} \frac{1}{n} \log \mathscr{L}_{t,x}^n \mathbb{1}(w_n) = \lim_{n\to\infty} \frac{1}{n} \log \lambda_{t,x}^n = \mathscr{E}P(t) \tag{9.10}$$

for all $x \in X_t'$. Fix $\lambda \in [0,1)$ and let $t = (t_1, t_2)$ and $t' = (t_1', t_2') \in \mathbb{R}^2$. Hölder's inequality implies that all the functions $\mathbb{R}^2 \ni t \mapsto \frac{1}{n} \log \mathscr{L}_{t,x}^n \mathbb{1}(w_n)$, $n \geq 1$, are convex. It thus follows from (9.10), that the function $\mathbb{R}^2 \ni t \mapsto \mathscr{E}P(t)$ is convex, whence continuous. Let

$$X_{\mathscr{E}}' = \bigcap_{t\in\mathbb{Q}^2} X_t'.$$

Since the set $\mathbb{Q}^2$ is countable, we have that $m(X_{\mathscr{E}}') = 1$. Along with (9.10), and density of $\mathbb{Q}^2$ in $\mathbb{R}^2$, the convexity of the functions $\mathbb{R}^2 \ni t \mapsto \frac{1}{n} \log \mathscr{L}_{t,x}^n \mathbb{1}(w_n)$ implies that for all $x \in X_{\mathscr{E}}'$ and all $t \in \mathbb{R}^2$, the limit $\lim_{n\to\infty} \frac{1}{n} \log \mathscr{L}_{t,x}^n \mathbb{1}(w_n)$ exists and represents a convex function, whence continuous. Since for all $t \in \mathbb{Q}^2$ this continuous function is equal to the continuous function $\mathscr{E}P$, we conclude that for all $x \in X_{\mathscr{E}}'$ and all $t \in \mathbb{R}^2$, we have

$$\lim_{n\to\infty} \frac{1}{n} \log \mathscr{L}_{t,x}^n \mathbb{1}(w_n) = \mathscr{E}P(t).$$

We are done.   □

Lemma 9.6 *Fix $t_2 \in \mathbb{R}$ and assume that there exist measurable functions $L : X \ni x \mapsto L_x \in \mathbb{R}$ and $c : X \ni x \mapsto c_x > 0$ such that*

$$S_n\varphi_{x,1}(z) \leq -nc_x + L_x \quad \text{for every } z \in \mathscr{J}_x \text{ and } n \geq 1. \tag{9.11}$$

Then the function $\mathbb{R} \ni t_1 \mapsto \mathscr{E}P(t_1, t_2) \in \mathbb{R}$ is strictly decreasing and

$$\lim_{t_1\to+\infty} \mathscr{E}P(t_1, t_2) = -\infty \quad \text{and} \quad \lim_{t_1\to-\infty} \mathscr{E}P(t_1, t_2) = +\infty \quad m\text{-a.e.} \tag{9.12}$$

Proof. Fix $x \in X_{\mathcal{E}}'$. Let $t_1 < t_1'$. Then by (9.11)

$$\sum_{z \in T_x^{-n}(w_n)} \exp(S_n \varphi_{(t_1,t_2)}(z))$$

$$= \sum_{z \in T_x^{-n}(w_n)} \exp(t_1 S_n \varphi_1(z)) \exp(t_2 S_n \varphi_2(z))$$

$$= \sum_{z \in T_x^{-n}(w_n)} \exp(t_1' S_n \varphi_1(z)) \exp(t_2 S_n \varphi_2(z)) \exp((t_1 - t_1') S_n \varphi_1(z))$$

$$\geq \sum_{z \in T_x^{-n}(w_n)} \exp(t_1' S_n \varphi_2(z)) \exp(t_2 S_n \varphi_2(z)) \exp((t_1 - t_1')(L_x - nc_x))$$

$$= \sum_{z \in T_x^{-n}(w_n)} \exp(S_n \varphi_{(t_1',t_2)}(z)) \exp((t_1' - t_1)(nc_x - L_x))$$

Therefore,

$$\frac{1}{n} \log \mathcal{L}_{t,x}^n \mathbb{1}(w_n) \geq \frac{1}{n} \log \mathcal{L}_{(t_1',t_2),x}^n \mathbb{1}(w_n) + (t_1' - t_1)(c_x - L_x/n).$$

Hence, letting $n \to \infty$, we get from Lemma 9.5 that $\mathscr{E}P(t_1,t_2) \geq \mathscr{E}P(t_1',t_2) + (t_1' - t_1)c_x$. It directly follows from this inequality that the function $t_1 \mapsto \mathscr{E}P(t_1,t_2)$ is strictly decreasing, that $\lim_{t_1 \to +\infty} \mathscr{E}P(t_1,t_2) = -\infty$ and that $\lim_{t_1 \to -\infty} \mathscr{E}P(t_1,t_2) = +\infty$. $\qquad\square$

9.2 Real Cones

We adapt the approach of Rugh [26] based on complex cones and establish real analyticity of the pressure function. Via Legendre transformation, this completes the proof of real analyticity of the multifractal spectrum (see Chap. 6).

Let $\mathscr{H}_x := \mathscr{H}_{\mathbb{R},x} := \mathscr{H}^\alpha(\mathcal{J}_x)$ and let $\mathscr{H}_{\mathbb{C},x} := \mathscr{H}_{\mathbb{R},x} \oplus i \mathscr{H}_{\mathbb{R},x}$ its complexification.

$$\mathscr{C}_x^s := \mathscr{C}_{\mathbb{R},x}^s := \{g \in \mathscr{H}_x : g(w_1) \leq e^{s Q_x \varrho^\alpha(w_1,w_2)} g(w_2) \text{ if } \varrho(w_1, w_2) \leq \xi\}. \tag{9.13}$$

Whenever it is clear what we mean by s, we also denote this cone by $\mathscr{C}_x$.

By $\mathscr{C}_x^+$ we denote the subset of all non-zero functions from $\mathscr{C}_x^s$. For $l \in (\mathscr{H}_x)^*$, the dual space of $\mathscr{H}_x$, we define

$$K(\mathscr{C}_x^s, l) := \sup_{g \in \mathscr{C}_x^+} \frac{\|l\|_\alpha \|g\|_\alpha}{|\langle l, g \rangle|}.$$

Then the *aperture* of $\mathscr{C}_x^s$ is

$$K(\mathscr{C}_x^s) := \inf\{K(\mathscr{C}_x^s, l) : l \in (\mathscr{H}_x)^*, l \neq 0\}.$$

Lemma 9.7 $K(\mathscr{C}_x^s) < \infty$. *This property of a cone is called an* outer regularity.

Proof. Let $w_k \in \mathscr{J}_x, k = 0, \ldots, N$ be such that $\bigcup_{k=1}^{L_x} B(w_k, \xi) = \mathscr{J}_x$. Define

$$l_0(g) := \sum_{k=1}^{L_x} g(w_k). \tag{9.14}$$

Then by Lemma 3.11 we have

$$\|g\|_\alpha \leq \left(sQ_x(\exp(sQ_x\xi^\alpha)) + 1\right)\|g\|_\infty$$

$$\leq \left(sQ_x(\exp(sQ_x\xi^\alpha)) + 1\right)\exp(sQ_x\xi^\alpha)l_0(g).$$

Note that $\|l_0\|_\alpha = L_x$, since $l_0(g) \leq L_x\|g\|_\infty \leq L_x\|g\|_\alpha$ and $l_0(\mathbb{1}) = L_x = L_x\|\mathbb{1}\|_\alpha$. Hence

$$\frac{\|l_0\|_\alpha\|g\|_\alpha}{\langle l_0, g\rangle} \leq K_x' := L_x\left(sQ_x(\exp(sQ_x\xi^\alpha)) + 1\right)\exp(sQ_x\xi^\alpha). \tag{9.15}$$

□

Let

$$s_x' := \frac{sQ_{x-1}\gamma_{x-1}^{-\alpha} + H_{x-1}\gamma_{x-1}^{-\alpha}}{Q_x}.$$

By (3.33) for $s > 1$, $s_x' < s$. Moreover, like in (3.32) we have the following.

Lemma 9.8 *Let* $g \in \mathscr{C}_x^s$ *and let* $w_1, w_2 \in \mathscr{J}_{\theta(x)}$ *with* $\varrho(w_1, w_2) \leq \xi$. *Then, for* $y \in T_x^{-1}(w_1)$

$$\frac{e^{\varphi(y)}}{e^{\varphi(T_y^{-1}(w_2))}}\frac{g(y)}{g(T_y^{-1}(w_2))} \leq \exp\left\{s_{\theta(x)}' Q_{\theta(x)}\varrho^\alpha(w_1, w_2)\right\}. \tag{9.16}$$

Consequently

$$\frac{\mathscr{L}_x g(w_1)}{\mathscr{L}_x g(w_2)} \leq \exp\left\{s_{\theta(x)}' Q_{\theta(x)}\varrho^\alpha(w_1, w_2)\right\}.$$

Lemma 9.9 *There is a measurable function* $C_R : X \to (0, \infty)$ *such that*

$$\frac{\mathscr{L}_{x-i}^i g(w)}{\mathscr{L}_{x-i}^i g(z)} \leq C_R(x) \quad \text{for every } i \geq j(x) \text{ and } g \in \mathscr{C}_x^s.$$

Proof. First, let $i = j(x)$. Let $a \in T_{x-i}^{-i}(z)$ be such that

$$e^{S_i \varphi(a)} g(a) = \sup_{y \in T_{x-i}^{-i}(z)} e^{S_i \varphi(y)} g(y).$$

By definition of $j(x)$, for any point $w \in \mathscr{J}_x$ there exists $b \in T_{x-i}^{-i}(w) \cap B(a, \xi)$. Therefore

$$\mathscr{L}_{x-i}^i g(w) \geq e^{S_i \varphi_{x-i}(b)} g(z) \geq \exp(S_i \varphi_{x-i}(b) - S_i \varphi_{x-i}(a)) e^{S_i \varphi_{x-i}(a)} e^{-sQ_x} g(a)$$

$$\geq \frac{\exp(-2\|S_{j(x)} \varphi_{x-j(x)}\|_\infty - sQ_x)}{\deg(T_{x-j}^j)} \mathscr{L}_{x-i}^i g(z) \geq (C_R(x))^{-1} \mathscr{L}_{x-i}^i g(z),$$

where

$$C_R(x) := \left(\frac{\exp\left(-sQ_x - 2\|S_{j(x)} \varphi_{x-j(x)}\|_\infty \right)}{\deg(T_{x-j}^{j(x)})} \right)^{-1} \geq 1. \tag{9.17}$$

The case $i > j(x)$ follows from the previous one, since $\mathscr{L}_{x-i}^{i-j(x)} g_{x-i} \in \mathscr{C}_{x-j(x)}^s$. □

Let $s > 1$ and $s' < s$. Define

$$\tau_x := \tau_{x,s,s'} := \sup_{r \in (0, \xi]} \frac{1 - \exp\left(-(s+s') Q_x r^\alpha \right)}{1 - \exp\left(-(s-s') Q_x r^\alpha \right)} \leq \frac{s+s'}{s-s'}. \tag{9.18}$$

Lemma 9.10 *For $g_x, f_x \in \mathscr{C}_x^{s'}$,*

$$\tau_x \frac{\sup_{y \in \mathscr{J}_x} |g_x(y)|}{\inf_{y \in \mathscr{J}_x} |f_x(y)|} f_x - g_x \in \mathscr{C}_{\mathbb{R},x}^s.$$

Proof. For all $w, z \in \mathscr{J}_x$ with $\varrho_x(z, w) < \xi$,

$$\tau_x \|g_x/f_x\|_\infty \left(\exp\left(sQ_x \varrho_x^\alpha(z, w) \right) f_x(z) - f_x(w) \right)$$

$$\geq \tau_x \|g_x/f_x\|_\infty \left(\exp\left(sQ_x \varrho_x^\alpha(z, w) \right) - \exp\left(s' Q_x \varrho_x^\alpha(z, w) \right) \right) f_x(z)$$

$$\geq \left(\exp\left(sQ_x \varrho_x^\alpha(z, w) \right) - \exp\left(-s' Q_x \varrho_x^\alpha(z, w) \right) \right) g_x(z)$$

$$\geq \exp\left(sQ_x \varrho_x^\alpha(z, w) \right) g_x(z) - g_x(w).$$

Then $\exp\left(sQ_x \varrho_x^\alpha(z, w) \right) \left(\tau_x \|g/f\|_\infty f_x(z) - g_x(z) \right) \geq \tau_x \|g/f\|_\infty f_x(w) - g_x(w).$ □

We say that $g_x \in \mathscr{C}_x^s$ is *balanced* if

$$\frac{f_x(y_1)}{f_x(y_2)} \le C_R(x) \quad \text{for all } y_1, y_2 \in \mathscr{J}_x. \tag{9.19}$$

Let $g_x, f_x \in \mathscr{C}_x^s$. Put $\beta_{x,s}(f_x, g_x) := \inf\{\tau > 0 : \tau f_x - g_x \in \mathscr{C}_x^s\}$ and define the *Hilbert projective distance* Pdist $: \mathscr{C}_x^s \times \mathscr{C}_x^s \to \mathbb{R}$ by the formula:

$$\text{Pdist}_x(f_x, g_x) := \text{Pdist}_{x,s}(f_x, g_x) := \log(\beta_{x,s}(f_x, g_x) \cdot \beta_{x,s}(g_x, f_x)).$$

Let

$$\Delta_x := \text{diam}_{\mathscr{C}_{x,\mathbb{R}}^s}(\mathscr{L}_{x-j}^j(\mathscr{C}_{x-j,\mathbb{R}}^s)),$$

where $\text{diam}_{\mathscr{C}_{x,\mathbb{R}}^s}$ is the diameter with respect to the projective distance and $j = j(x)$. Then by Lemmas 9.8, 9.9 and 9.10 we get the following.

Lemma 9.11 *If $g_x, f_x \in \mathscr{C}_x^{s'}$ are balanced, then*

$$\text{Pdist}_x(f_x, g_x) \le 2 \log\left(\frac{s + s'}{s - s'} \cdot C_R(x)\right)$$

and, consequently,

$$\Delta_x \le 2 \log\left(\frac{s + s'}{s - s'} \cdot C_R(x)\right).$$

9.3   Canonical Complexification

Following the ideas of Rugh [26] we now extend real cones to complex ones. Define $\mathscr{C}_x^* := \{l \in (\mathscr{H}_x)^* : l|_{\mathscr{C}_x} \ge 0\}$ and

$$\mathscr{C}_{\mathbb{C},x}^s := \{g \in \mathscr{H}_{\mathbb{C},x} : \forall_{l_1, l_2 \in \mathscr{C}_x^*} \text{Re}\langle l_1, g\rangle \overline{\langle l_2, g\rangle} \ge 0\}.$$

Denote also by $\mathscr{C}_{\mathbb{C},x}^+$ the set of all $g \in \mathscr{C}_{\mathbb{C},x}^s$ such that $g \not\equiv 0$. There are other equivalent definitions of $\mathscr{C}_{\mathbb{C},x}^s$. The first one is called *polarization identity* by Rugh in [26, Proposition 5.2].

Proposition 9.12 (Polarization identity)

$$\mathscr{C}_{\mathbb{C},x}^s = \{a(f^* + ig^*) : f^* \pm g^* \in \mathscr{C}_{\mathbb{R},x}^+ \text{ and } a \in \mathbb{C}\}.$$

In our case we can also define $\mathscr{C}_{\mathbb{C},x}^s$ as follows. Let $\varrho(w, w') < \xi$. Define

$$l_{w,w'}(g) := g(w) - e^{-sQ_x\varrho^\alpha(w,w')}g(w')$$

and

$$F_x := \{l_{w,w'} : \varrho(w, w') < \xi\} \subset \mathscr{C}_x^*.$$

Then

$$\mathscr{C}_x^s = \{g \in \mathscr{H}_x : \forall_{l \in F_x} l(g) \geq 0\}.$$

Later in this section we use the following two facts about geometry of complex numbers. The first one is obvious and the second is Lemma 9.3 from [26].

Lemma 9.13 *Given* $c_1, c_2 > 0$ *there exist* $p_1, p_2 > 0$ *such that if* $s_0 := c_1 p_2$ *and*

$$Z \in \{re^{iu} : 1 \leq 1 + s_0^2, |u| \leq 2p_1 + 2s_0\},$$

then there exist $\alpha, \beta, \gamma > 0$ *such that* $\mathrm{Re}\, Z \geq \alpha$, $\mathrm{Re}\, Z \leq \beta$, $\mathrm{Im}\, Z \leq \gamma$ *and* $\gamma c_2 < \alpha$.

Lemma 9.14 *Let* $z_1, z_2 \in \mathbb{C}$ *be such that* $\mathrm{Re}\, z_1 > \mathrm{Re}\, z_2$ *and define* $u \in \mathbb{C}$ *though*

$$e^{i\,\mathrm{Im}\, z_1} u \equiv \frac{e^{z_1} - e^{z_2}}{e^{\mathrm{Re}\, z_1} - e^{\mathrm{Re}\, z_2}}.$$

Then

$$|\mathrm{Arg}\, u| \leq \frac{|\mathrm{Im}(z_1 - z_2)|}{\mathrm{Re}(z_1 - z_2)} \quad \text{and} \quad 1 \leq |u^2| \leq 1 + \left(\frac{\mathrm{Im}(z_1 - z_2)}{\mathrm{Re}(z_1 - z_2)}\right)^2.$$

Let $\varphi = \mathrm{Re}\, \varphi + i\,\mathrm{Im}\, \varphi$ be such that $\mathrm{Re}\, \varphi, \mathrm{Im}\, \varphi \in \mathscr{H}^\alpha(\mathscr{J})$. We now consider the corresponding complex Perron–Frobenius operators $\mathscr{L}_{x,\varphi}$ defined by

$$\mathscr{L}_{x,\varphi} g_x(w) = \sum_{T_x(z)=w} e^{\varphi_x(z)} g_x(z), \quad w \in \mathscr{J}_{\theta(x)}.$$

Lemma 9.15 *Let* $w, w', z, z' \in \mathscr{J}_x$ *such that* $\varrho(w, w') < \xi$ *and* $\varrho(z, z') < \xi$. *Then,* *for all* $g_1, g_2 \in \mathscr{C}_{x,\mathbb{R}}^s$,

$$\frac{l_{w,w'}(\mathscr{L}_{x,\varphi} g_1)\overline{l_{z,z'}(\mathscr{L}_{x,\varphi} g_2)}}{l_{w,w'}(\mathscr{L}_{x,\mathrm{Re}\,\varphi} g_1)l_{z,z'}(\mathscr{L}_{x,\mathrm{Re}\,\varphi} g_2)} = Z,$$

where

$$Z \in A_x := \{re^{iu} : 1 \leq r \leq 1 + s_0^2, |u| \leq 2\|\mathrm{Im}\,\varphi\|_\infty + 2s_0\} \tag{9.20}$$

and

$$s_0 := \frac{v_\alpha(\mathrm{Im}\,\varphi)\gamma_x^{-\alpha}}{(s - s_{\theta(x)}')Q_{\theta(x)}}. \tag{9.21}$$

Proof. For $y \in T_x^{-1}(w)$, by y' we denote $T_y^{-1}(w')$. Then for $g \in \mathscr{C}_x$

$$l_{w,w'}(\mathscr{L}_{x,\varphi}g) := \mathscr{L}_{x,\varphi}g(w) - e^{-sQ_x\varrho^\alpha(w,w')}\mathscr{L}_{x,\varphi}g(w')$$

$$= \sum_{y \in T_x^{-1}(w)} e^{\varphi(y)}g(y) - e^{-sQ_x\varrho^\alpha(w,w')}e^{\varphi(y')}g(y') = \sum_{y \in T_x^{-1}(w)} n_y(\varphi, g),$$

where

$$n_y(\varphi, g) := e^{\varphi(y)}g(y) - e^{-sQ_x\varrho^\alpha(w,w')}e^{\varphi(y')}g(y').$$

Define implicitly u_y so that $n_y(\mathrm{Re}\,\varphi, g)e^{i\,\mathrm{Im}\,\varphi(y)}u_y = n_y(\varphi, g)$. Put $z_1 := \varphi(y) + \log g(y)$ and $z_2 := -sQ_x\varrho^\alpha(w, w') + \varphi(y') + \log g(y')$. Then

$$e^{i\,\mathrm{Im}\,z_1}u_y = \frac{e^{z_1} - e^{z_2}}{e^{\mathrm{Re}\,z_1} - e^{\mathrm{Re}\,z_2}}.$$

By (9.16)

$$\mathrm{Re}\,\varphi(y) - \log g(y) - (\mathrm{Re}\,\varphi(y') + \log g(y')) \geq -s'_{\theta(x)}Q_{\theta(x)}\varrho^\alpha(w_1, w_2).$$

Hence

$$\mathrm{Re}(z_1 - z_2) \geq (s - s'_{\theta(x)})Q_{\theta(x)}\varrho^\alpha(w_1, w_2).$$

We also have that

$$|\mathrm{Im}(z_1 - z_2)| \leq v_\alpha(\mathrm{Im}\,\varphi)\gamma_x^{-\alpha}\varrho^\alpha(w_1, w_2),$$

since $\mathrm{Im}(z_1 - z_2) = \mathrm{Im}\,\varphi(y) - \mathrm{Im}\,\varphi(y')$. Therefore, by Lemma 9.14

$$|\mathrm{Arg}\,u_y| \leq s_0 := \frac{v_\alpha(\mathrm{Im}\,\varphi)\gamma_x^{-\alpha}}{(s - s'_{\theta(x)})Q_{\theta(x)}} \quad \text{and} \quad 1 \leq |u_y|^2 \leq 1 + s_0^2.$$

Since

$$l_{w,w'}(\mathscr{L}_{x,\varphi}g) = \sum_{y \in T_x^{-1}(w)} n_y(\varphi, g) = \sum_{y \in T_x^{-1}(w)} e^{i\,\mathrm{Im}\,\varphi(y)}u_y n_y(\mathrm{Re}\,\varphi, g),$$

$$\frac{l_{w,w'}(\mathscr{L}_{x,\varphi}g)}{l_{w,w'}(\mathscr{L}_{x,\mathrm{Re}\,\varphi}g)} = Z,$$

where

$$Z \in A_x := \{re^{iu} : 1 \leq r \leq 1 + s_0^2, |u| \leq 2||\,\mathrm{Im}\,\varphi||_\infty + 2s_0\}.$$

Similarly

$$\frac{l_{w,w'}(\mathscr{L}_{x,\varphi}g_1)\overline{l_{z,z'}}(\mathscr{L}_{x,\varphi}g_2)}{l_{w,w'}(\mathscr{L}_{x,\mathrm{Re}\,\varphi}g_1)l_{z,z'}(\mathscr{L}_{x,\mathrm{Re}\,\varphi}g_2)} = Z$$

for possibly another $Z \in A_x$.   □

Let p_1, p_2 be the real numbers given by Lemma 9.13 with

$$c_1 = \frac{\gamma_x^{-\alpha}}{(s - s_x')Q_x} \quad \text{and} \quad c_2 = \cosh\frac{\varDelta_x}{2}.$$

Having Lemmas 9.15, 9.13 and 9.11 the following proposition is a consequence of the proof of Theorem 6.3 in [26].

Proposition 9.16 *Let $j = j(x)$. If*

$$\| \mathrm{Im}\, S_j\varphi_{x-j} \|_\infty \le p_1 \quad and \quad v_\alpha(\mathrm{Im}\, S_j\varphi_{x-j}) \le p_2, \tag{9.22}$$

then

$$\mathscr{L}_{x-j}^j(\mathscr{C}_{\mathbb{C},x-j}^s) \subset \mathscr{C}_{\mathbb{C},x}^s.$$

Let l_0 (the functional defined by (9.14)). Then by Lemma 5.3 in [26] we get

$$K := K(\mathscr{C}_{\mathbb{C},x}^s, l_0) := \sup_{g \in \mathscr{C}_{\mathbb{C},x}^+} \frac{\|l_0\|_\alpha \|g\|_\alpha}{|\langle l_0, g \rangle|} \le K_x := 2\sqrt{2}K_x',$$

where K_x' is defined by (9.15). By l we denote the functional which is a normalized version of $(1/L_x)l_0$. So $\|l\|_\alpha = 1$. Then, for every $g \in \mathscr{C}_{\mathbb{C},x}^s$,

$$1 \le \frac{\|g\|_\alpha}{\langle l, g \rangle} \le K_x. \tag{9.23}$$

9.4 The Pressure is Real-Analytic

We are now in position to prove the main result of this chapter. Here, we assume that $T : \mathscr{J} \to \mathscr{J}$ is uniformly expanding random map. Then there exists $j \in \mathbb{N}$ such that $j(x) = j$ for all $x \in X$. Without loss of generality we assume that $j = 1$.

Theorem 9.17 *Let $t_0 = (t_1, \ldots, t_n) \in \mathbb{R}^n$, $R > 0$ and let*

$$D(t_0, R) := \{z = (z_1, \ldots, z_n) \in \mathbb{C}^n : \forall_k |z_k - t_k| < R\}.$$

Assume that the following conditions are satisfied.

(a) *For every $x \in X$ and every $w \in \mathscr{J}_x$, $z \mapsto \varphi_{z,x}(w)$ is holomorphic on $D(t_0, R)$.*
(b) *For $z \in \mathbb{R}^n \cap D(t_0, R)$, $\varphi_{z,x} \in \mathscr{H}_{\mathbb{R},x}$.*

(c) *For all $z \in D(t_0, R)$ and all $x \in X$, there exists H such that $\|\varphi_{z,x}\|_\alpha \leq H$.*

(d) *For every $\varepsilon > 0$ there exists $\delta > 0$ such that for all $z \in D(t_0, \delta)$ and all $x \in X$,*

$$\| \operatorname{Im} \varphi_{z,x} \|_\alpha \leq \varepsilon.$$

Then the function $D(t_0, R) \cap \mathbb{R}^n \ni z \mapsto \mathscr{E}P(\varphi_z)$ is real-analytic.

Proof. Since we assume that the measurable constants are uniform for $x \in X$ we get that from Proposition 9.16 and condition (d) that there exists $r > 0$ such that, for all $z \in D(t_0, r)$ and all $x \in X$,

$$\mathscr{L}_{z,x-1}(\mathscr{C}^s_{\mathbb{C},x-1}) \subset \mathscr{C}^s_{\mathbb{C},x}.$$

Then by (9.23),

$$\frac{\|\mathscr{L}^n_{z,x-n}(\mathbb{1})\|_\alpha}{l_x(\mathscr{L}^n_{z,x-n}(\mathbb{1}))} \leq K.$$

Therefore, by Montel Theorem, the family $\frac{\mathscr{L}^n_{z,x-n}(\mathbb{1})(w)}{l_x(\mathscr{L}^n_{z,x-n}(\mathbb{1}))}$ is normal. Since, for all $z \in \mathbb{R}^n \cap D(t_0, r)$ and all $x \in X$ we have that

$$\frac{\mathscr{L}^n_{z,x-n}(\mathbb{1})(w)}{l_x(\mathscr{L}^n_{z,x-n}(\mathbb{1}))} \xrightarrow[n \to \infty]{} \frac{q_{z,x}(w)}{l_x(q_{z,x})},$$

we conclude that there exists an analytic function $z \mapsto g_{z,x}(w)$ such that

$$\frac{\mathscr{L}^n_{z,x-n}(\mathbb{1})(w)}{l_x(\mathscr{L}^n_{z,x-n}(\mathbb{1}))} \xrightarrow[n \to \infty]{} g_{z,x}(w). \tag{9.24}$$

Since, in addition,

$$\mathscr{L}_x\left(\frac{\mathscr{L}^n_{z,x-n}(\mathbb{1})(w)}{l_x(\mathscr{L}^n_{z,x-n}(\mathbb{1}))}\right) = \frac{\mathscr{L}^{n+1}_{z,x-n}(\mathbb{1})(w)}{l_{x_1}(\mathscr{L}^{n+1}_{z,x-n}(\mathbb{1}))} \cdot l_{x_1}\left(\mathscr{L}_{z,x}\left(\frac{\mathscr{L}^n_{z,x-n}(\mathbb{1})(w)}{l_x(\mathscr{L}^n_{z,x-n}(\mathbb{1}))}\right)\right),$$

we therefore get that

$$\mathscr{L}_x\left(\frac{\mathscr{L}^n_{z,x-n}(\mathbb{1})(w)}{l_x(\mathscr{L}^n_{z,x-n}(\mathbb{1}))}\right) \xrightarrow[n \to \infty]{} l_{x_1}(\mathscr{L}_x(g_{z,x}))g_{x_1,z}.$$

Thus, using again (9.24), we obtain $\mathscr{L}_{z,x}(g_{z,x}) = l_{x_1}(\mathscr{L}_{z,x}(g_{z,x}))g_{x_1,z}$. As for all $z \in D(t_0, r) \cap \mathbb{R}^n$,

$$g_{z,x} = \frac{q_{z,x}}{l_x(q_{z,x})} = \frac{L_x q_{z,x}}{\sum_{k=0}^N q_{z,x}(w_k)},$$

we conclude that,

$$l_{x_1}(\mathscr{L}_{z,x} g_{z,x}) = l_{x_1}(\mathscr{L}_{z,x} \frac{q_{z,x}}{l_x(q_{z,x})}) = \lambda_{z,x} \frac{l_{x_1}(q_{x_1,z})}{l_x(q_{z,x})}. \tag{9.25}$$

By the very definitions

$$l_{x_1}(\mathscr{L}_{z,x} g_{z,x}) = (1/L_x) \sum_{k=1}^{L_x} \mathscr{L}_{z,x} g_{z,x}(w_k)$$

and

$$\mathscr{L}_{z,x} g_{z,x}(w) = \sum_{y \in T_x^{-1}(w)} e^{\varphi_{z,x}(y)} g_{z,x}(y).$$

Denote $g_{z,x}(w)$ by $F(z)$ and $\varphi_{z,x}(w)$ by $G(z)$. Then, for $z = (z_1, \ldots, z_n) \in D(t_0, r/2)$, and $\Gamma(u) = z + ((r/2)e^{2\pi i u_1}, \ldots, (r/2)e^{2\pi i u_n})$, where $u = (u_1, \ldots, u_n) \in [0, 2\pi]^n$, by the Cauchy Integral Formula,

$$\left| \frac{\partial F}{\partial z_k}(z) \right| = \left| \frac{1}{(2\pi i)^2} \int_\Gamma \frac{F(\xi)}{(\xi_1 - z_1) \ldots (\xi_k - z_k)^2 \ldots (\xi_2 - z_2)} d\xi \right| \le 2K/r$$

for $k = 1, \ldots, n$. Similarly we obtain that

$$\left| \frac{\partial G}{\partial z_k}(z) \right| \le 2H/r$$

for $k = 1, \ldots, n$. Then, for $k = 1, \ldots, n$,

$$\left| \frac{\partial e^{\varphi_{z,x}(y)} g_{z,x}(y)}{\partial z_k} \right| = \left| \frac{\partial \varphi_{z,t}(w)}{\partial z_k} e^{\varphi_{z,x}(y)} g_{z,x}(y) + e^{\varphi_{z,x}(y)} \frac{\partial g_{z,x}(y)}{\partial z_k} \right|$$

$$\le (2H/r)e^H K + e^H (2K/r).$$

It follows that there exists C_g such that for all $x \in X$,

$$\left| \frac{\partial l_{x_1}(\mathscr{L}_{z,x} g_{z,x})}{\partial z_k} \right| \le C_g. \tag{9.26}$$

Using (3.19) we obtain that

$$C_\varphi^{-1} \le q_{t_0,x}(y) \le C_\varphi,$$

and then

$$C_\varphi^{-1} \le l_x(q_{t_0,x}(y)) \le C_\varphi$$

for all $x \in X$. Moreover, it follows from Lemma 3.6 that $\lambda_{t_0,x} \geq \exp(-\|\varphi_{t_0,x}\|_\infty)$. Then

$$z_0 := l_{x_1}(\mathscr{L}_{t_0,x} g_{t_0,x}) = \lambda_{t,x} \frac{l_{x_1}(q_{t,x_1})}{l_x(q_{t,x})} \geq \exp(-\sup_{x \in X}\|\varphi_x\|_\infty)C_\varphi^{-2} > 0.$$

Hence, by (9.26), there exists $r_1 > 0$ so small that

$$l_{x_1}(\mathscr{L}_{z,x} g_{z,x}) \in D(z_0, z_0/2)$$

for all $z \in D(t_0, r_1)$. Therefore, for all $x \in X$ we can define the function

$$D(t_0, r_1) \ni z \mapsto \log l_{x_1}(\mathscr{L}_{z,x} g_{z,x}) \in \mathbb{C}.$$

Now consider the holomorphic function

$$z \mapsto \int \log l_{x_1}(\mathscr{L}_{z,x} g_{z,x}) dm(x).$$

Since the measure m is θ-invariant, by (9.25)

$$\int \log l_{x_1}(\mathscr{L}_{z,x} g_{z,x}) dm(x) = \int \log \lambda_{z,x} \frac{l_{x_1}(q_{z,x_1})}{l_x(q_{z,x})} dm(x)$$

$$= \int \log \lambda_{z,x} dm + \int l_{x_1}(q_{z,x_1}) dm - \int l_x(q_{z,x}) dm(x)$$

$$= \int \log \lambda_{z,x} dm = \mathscr{E}P(\varphi_t)$$

for $z \in D(t_0, r_1) \cap \mathbb{R}^n$. Therefore the function $D(t_0, r_1) \cap \mathbb{R}^n \ni z \mapsto \mathscr{E}P(\varphi_z)$ is real-analytic. $\qquad\Box$

9.5   Derivative of the Pressure

Now, let $T : \mathscr{J} \to \mathscr{J}$ be uniformly expanding random map. Throughout the section, we assume that $\varphi \in \mathscr{H}_m(\mathscr{J})$ is a potential such that there exist measurable functions $L : X \ni x \mapsto L_x \in \mathbb{R}$ and $c : X \ni x \mapsto c_x > 0$ such that

$$S_n \varphi_x(z) \leq -nc_x + L_x \tag{9.27}$$

for every $z \in \mathscr{J}_x$ and n and $\psi \in \mathscr{H}_m(\mathscr{J})$. For $t \in \mathbb{R}$, define

$$\varphi_t := t\varphi + \psi.$$

Let $R > 0$ and let $|t_0| \leq R/2$. Since we are in the uniform case, it follows from Remark 9.2 that there exist constants A_R and B_R such that, for $t \in [-R, R]$,

$$\left\| \frac{\tilde{\mathscr{L}}_{t,x}^n g_x}{q_{\theta^n}(x)} - \left(\int g_x dv_{t,x} \right) \right\|_\infty \leq \left(\|g_x\|_\infty + 2\frac{v(g_x)}{Q} \right) A_R B_R^n. \tag{9.28}$$

Proposition 9.18

$$\frac{d\mathscr{E}P(t)}{dt} = \int \varphi_x d\mu_x^t dm(x) = \int \varphi d\mu^t.$$

Proof. Assume without loss of generality that $|t| \leq R/2$ for some $R > 0$. Let $x \mapsto y(x) \in Y_x$ be a measurable function and let

$$\mathscr{E}P(t, n) := \int \frac{1}{n} \log \mathscr{L}_{t,x}^n \mathbb{1}_x(y(x_n)) dm(x).$$

Then $\lim_{n \to \infty} \mathscr{E}P(t, n) = \mathscr{E}P(t)$ by Lemma 4.6. Fix $x \in X$ and put $y_n := y(x_n)$. Observe that

$$\frac{d\mathscr{L}_{t,x}^n \mathbb{1}_x(y_n)}{dt} = \sum_{y \in T_x^{-n}(y_n)} e^{S_n(\varphi_x^t)(y)} S_n \varphi_x(y)$$

$$= \sum_{j=0}^{n-1} \sum_{y \in T_x^{-n}(y_n)} e^{S_n(\varphi_x^t)(y)} \varphi_{x_j}(T_x^j y) = \sum_{j=0}^{n-1} \mathscr{L}_{t,x}^n (\varphi_{x_j} \circ T_x^j)(y_n).$$

Since $S_n(\varphi_x^t)(y) = S_j(\varphi_x^t)(y) + S_{n-j}(\varphi_{x_j}^t)(T_x^j y)$ we have that

$$\mathscr{L}_{t,x}^n (\varphi_{x_j} \circ T_x^j)(y(x_n)) = \mathscr{L}_{t,x_j}^{n-j} (\varphi_{x_j} \mathscr{L}_{t,x}^j \mathbb{1}_x)(y(x_n)).$$

Then by a version of Leibniz integral rule (see for example [23], Proposition 7.8.4, p. 40)

$$\frac{d\mathscr{E}P(t, n)}{dt} = \int \frac{\frac{1}{n} \sum_{j=0}^{n-1} \mathscr{L}_{x_j,t}^{n-j} (\varphi_{x_j} \mathscr{L}_{t,x}^j \mathbb{1}_x)(y(x_n))}{\mathscr{L}_{t,x}^n \mathbb{1}_x(y_n)} dm(x).$$

Since

$$\mathscr{L}_{t,x_j}^{n-j} (\varphi_{x_j} \mathscr{L}_{t,x}^j \mathbb{1})(y_n) = \lambda_x^n \tilde{\mathscr{L}}_{t,x_j}^{n-j} \left(\varphi_{t,x_j} \tilde{\mathscr{L}}_{t,x}^j \mathbb{1}_x \right)(y_n)$$

and

$$\mathscr{L}_{t,x}^n \mathbb{1}_x(y_n) = \lambda_x^n \tilde{\mathscr{L}}_{t,x}^n \mathbb{1}_x(y_n),$$

we have that

$$\frac{\mathscr{L}_{t,x}^n(\varphi_{x_j} \circ T_x^j)(y_n)}{\mathscr{L}_{t,x}^n \mathbb{1}_x(y_n)} = \frac{\tilde{\mathscr{L}}_{t,x_j}^{n-j}\left(\varphi_{x_j}\tilde{\mathscr{L}}_{t,x}^j \mathbb{1}_x\right)(y_n)}{\tilde{\mathscr{L}}_{t,x}^n \mathbb{1}_x(y_n)}. \tag{9.29}$$

The function $\varphi_{x_j}\tilde{\mathscr{L}}_{t,x}^j \mathbb{1}_x$ is uniformly bounded. So does its Hölder variation. Therefore it follows from (9.28), that there exists a constant A_R and B_R such that

$$\left\|\tilde{\mathscr{L}}_{t,x_j}^{n-j}\left(\varphi_{x_j}\tilde{\mathscr{L}}_{t,x}^j \mathbb{1}_x\right)(y_n)/q_{x_n} - \left(\int \varphi_{x_j}\tilde{\mathscr{L}}_{t,x}^j \mathbb{1}_x dv_{x_j}^t\right)\right\|_\infty \le A_R B_R^{n-j}$$

and

$$\left\|\tilde{\mathscr{L}}_{t,x}^n(\mathbb{1}_x)(y_n)/q_{x_n} - \mathbb{1}_{x_n}\right\|_\infty \le A_R B_R^n,$$

From this by (9.29) it follows that

$$\frac{\int \varphi_{x_j}\tilde{\mathscr{L}}_{t,x}^j \mathbb{1}_x dv_{x_j}^t - A_R B_R^{n-j}}{1 + A_R B_R^n} \le \frac{\mathscr{L}_{t,x}^n(\varphi_{x_j} \circ T_x^j)(y_n)}{\mathscr{L}_x^n \mathbb{1}_{Y_x}(y_n)} \le \frac{\int \varphi_{x_j}\tilde{\mathscr{L}}_{t,x}^j \mathbb{1}_x dv_{x_j}^t + A_R B_R^{n-j}}{1 - A_R B_R^n}.$$

Since m is θ-invariant, we have that

$$\int\int \varphi_{x_j}\tilde{\mathscr{L}}_{t,x}y^j \mathbb{1}_x dv_{x_j}^t \, dm(x) = \int\int \varphi_x \tilde{\mathscr{L}}_{x-j,t}^j \mathbb{1}_{x-j} dv_x^t dm(x).$$

Hence, for large n,

$$\frac{\int\int \varphi_x\left(\frac{1}{n}\sum_{j=0}^{n-1}\tilde{\mathscr{L}}_{x-j,t}^j \mathbb{1}_{x-j}\right)dv_x^t dm(x) - \frac{1}{n}\sum_{j=0}^{n-1}(A_R B_R^{n-j})}{1 + A_R B_R^n} \le \frac{d\mathscr{E}P(\varphi^t, n)}{dt}$$

$$\le \frac{\int\int \varphi_x\left(\frac{1}{n}\sum_{j=0}^{n-1}\tilde{\mathscr{L}}_{x-j,t}^j \mathbb{1}_{x-j}\right)dv_x^t dm(x) - \frac{1}{n}\sum_{j=0}^{n-1}(A_R B_R^{n-j})}{1 - A_R B_R^n}.$$

Therefore

$$\lim_{n\to\infty} \frac{d\mathscr{E}P(t, n)}{dt} = \int \varphi_x d\mu_x^t dm(x)$$

uniformly for $t \in [-R, R]$. $\qquad\square$

References

1. Arnold, L.: Random Dynamical Systems. Springer Monographs in Mathematics. Springer, Berlin (1998)
2. Arnold, L., Evstigneev, I.V., Gundlach, V.M.: Convex-valued random dynamical systems: a variational principle for equilibrium states. Random Oper. Stochast. Equat. **7**(1), 23–38 (1999). doi: 10.1515/rose.1999.7.1.23. http://dx.doi.org/10.1515/rose.1999.7.1.23
3. Bahnmüller, J., Bogenschütz, T.: A Margulis–Ruelle inequality for random dynamical systems. Arch. Math. (Basel) **64**(3), 246–253 (1995). doi: 10.1007/BF01188575. http://dx.doi.org/10.1007/BF01188575
4. Bogenschütz, T.: Entropy, pressure, and a variational principle for random dynamical systems. Random Comput. Dyn. **1**(1), 99–116 (1992/93)
5. Bogenschütz, T., Gundlach, V.M.: Ruelle's transfer operator for random subshifts of finite type. Ergod. Theor. Dyn. Syst. **15**(3), 413–447 (1995)
6. Bogenschütz, T., Ochs, G.: The Hausdorff dimension of conformal repellers under random perturbation. Nonlinearity **12**(5), 1323–1338 (1999)
7. Bowen, R.: Equilibrium States and the Ergodic Theory of Anosov Diffeomorphisms. Lecture Notes in Mathematics, vol. 470. Springer, Berlin (1975)
8. Brück, R.: Geometric properties of Julia sets of the composition of polynomials of the form $z^2 + c_n$. Pac. J. Math. **198**(2), 347–372 (2001)
9. Brück, R., Büger, M.: Generalized iteration. Comput. Meth. Funct. Theor. **3**(1–2), 201–252 (2003)
10. Crauel, H.: Random Probability Measures on Polish Spaces. Stochastics Monographs, vol. 11. Taylor & Francis, London (2002)
11. Crauel, H., Flandoli, F.: Hausdorff dimension of invariant sets for random dynamical systems. J. Dyn. Differ. Equat. **10**(3), 449–474 (1998). doi: 10.1023/A:1022605313961. http://dx.doi.org/10.1023/A:1022605313961
12. Denker, M., Gordin, M.: Gibbs measures for fibred systems. Adv. Math. **148**(2), 161–192 (1999)
13. Denker, M., Urbański, M.: On the existence of conformal measures. Trans. Am. Math. Soc. **328**(2), 563–587 (1991)
14. Deschamps, V.M.: Equilibrium states for non-Hölderian random dynamical systems. Random Comput. Dyn. **5**(4), 319–335 (1997)
15. Falconer, K.: Techniques in Fractal Geometry. Wiley, Chichester (1997)
16. Ibragimov, I.A., Linnik, Y.V.: Independent and Stationary Sequences of Random Variables. Wolters-Noordhoff Publishing, Groningen (1971). With a supplementary chapter by I. A. Ibragimov and V. V. Petrov, Translation from the Russian edited by J. F. C. Kingman

17. Kifer, Y.: Equilibrium states for random expanding transformations. Random Comput. Dyn. **1**(1), 1–31 (1992/93)
18. Kifer, Y.: Fractal dimensions and random transformations. Trans. Am. Math. Soc. **348**(5), 2003–2038 (1996)
19. Kifer, Y.: Thermodynamic formalism for random transformations revisited. Stochast. Dyn. **8**, 77–102 (2008)
20. Kifer, Y., Liu, P.D.: Random dynamics. In: Handbook of Dynamical Systems, vol. 1B, pp. 379–499. Elsevier B.V., Amsterdam (2006)
21. Liu, P.D.: Entropy formula of Pesin type for noninvertible random dynamical systems. Math. Z. **230**(2), 201–239 (1999). doi: 10.1007/PL00004694. http://dx.doi.org/10.1007/PL00004694
22. Liu, P.D., Qian, M.: Smooth Ergodic Theory of Random Dynamical Systems. Lecture Notes in Mathematics, vol. 1606. Springer, Berlin (1995)
23. Malliavin, P.: Integration and Probability, Graduate Texts in Mathematics, vol. 157. Springer, New York (1995). With the collaboration of Hélène Airault, Leslie Kay and Gérard Letac, Edited and translated from the French by Kay, With a foreword by Mark Pinsky
24. Przytycki, F., Urbański, M.: Conformal Fractals: Ergodic Theory Methods. London Mathematical Society, Lecture Note Series 371. Cambridge University Press, Cambridge (2010)
25. Ruelle, D.: Thermodynamic Formalism. Encyclopedia of Mathematics and its Applications, vol. 5. Addison-Wesley Publishing Co., Reading, Mass (1978). The mathematical structures of classical equilibrium statistical mechanics, With a foreword by Giovanni Gallavotti and Gian-Carlo Rota
26. Rugh, H.H.: Cones and gauges in complex spaces: spectral gaps and complex Perron–Frobenius theory. Ann. Math. **171**(3), 1707–1752 (2010)
27. Rugh, H.H.: On the dimension of conformal repellors. randomness and parameter dependency. Preprint 2005, Ann. Math. **168**(3), 695–748 (2008)
28. Viana, M.: Stochastic dynamics of deterministic systems. Lecture Notes XXI Brazilian Mathematics Colloquium, IMPA, Rio de Janeiro (1997)
29. Walters, P.: Invariant measures and equilibrium states for some mappings which expand distances. Trans. Am. Math. Soc. **236**, 121–153 (1978)

Index

aperture, 98

backward visiting sequence, 11
balanced, 100
base map, 8
Bowen's Formula, 48
Brück and Büger polynomial systems, 83

classical conformal expanding random
    systems, 80
classical expanding random system, 76
concave Legendre transform, 58
conformal
    DG*-systems, 87
    expanding random map, 47
    uniformly expanding map, 47

DG*-system, 86
DG-system, 84

essential, 12
essentially random, 51
exhaustively visiting way, 12
expanding in the mean, 69
expanding random map, 8
expected pressure, 42

Gibbs family, 18
Gibbs property, 35

Hölder continuous with an exponent α, 13, 71

Hilbert projective distance, 100

induced map, 70
induced potential, 72
integrability of the logarithm of the transfer
    operator, 41
invariant density, 23

measurability
    of the degree, 9
    of cardinality of covers, 18
    of the transfer operator, 41
measurable expanding random map, 40
measurably expanding, 8

outer regularity, 98

polarization identity, 100
pressure function, 33
projective distance, 100
pseudo-pressure function, 18

random cantor set, 54
random compact subsets of Polish spaces, 43
random repeller, 75
random Sierpiński gasket, 5
repeller over U, 75
RPF-theorem, 17

T-invariance
    of a family of measures, 40

V. Mayer et al., *Distance Expanding Random Mappings, Thermodynamical Formalism,*
Gibbs Measures and Fractal Geometry, Lecture Notes in Mathematics 2036,
DOI 10.1007/978-3-642-23650-1, © Springer-Verlag Berlin Heidelberg 2011

of a family of measures, 17
of a measure, 40
temperature function, 58
topological exactness, 9
transfer
    dual operators, 19
    operator, 13

uniform openness, 8
uniformly expanding random map, 9

visiting sequence, 11
visiting way, 12

LECTURE NOTES IN MATHEMATICS  Springer

Edited by J.-M. Morel, B. Teissier; P.K. Maini

Editorial Policy (for the publication of monographs)

1. Lecture Notes aim to report new developments in all areas of mathematics and their applications - quickly, informally and at a high level. Mathematical texts analysing new developments in modelling and numerical simulation are welcome.

   Monograph manuscripts should be reasonably self-contained and rounded off. Thus they may, and often will, present not only results of the author but also related work by other people. They may be based on specialised lecture courses. Furthermore, the manuscripts should provide sufficient motivation, examples and applications. This clearly distinguishes Lecture Notes from journal articles or technical reports which normally are very concise. Articles intended for a journal but too long to be accepted by most journals, usually do not have this "lecture notes" character. For similar reasons it is unusual for doctoral theses to be accepted for the Lecture Notes series, though habilitation theses may be appropriate.

2. Manuscripts should be submitted either online at www.editorialmanager.com/lnm to Springer's mathematics editorial in Heidelberg, or to one of the series editors. In general, manuscripts will be sent out to 2 external referees for evaluation. If a decision cannot yet be reached on the basis of the first 2 reports, further referees may be contacted: The author will be informed of this. A final decision to publish can be made only on the basis of the complete manuscript, however a refereeing process leading to a preliminary decision can be based on a pre-final or incomplete manuscript. The strict minimum amount of material that will be considered should include a detailed outline describing the planned contents of each chapter, a bibliography and several sample chapters.

   Authors should be aware that incomplete or insufficiently close to final manuscripts almost always result in longer refereeing times and nevertheless unclear referees' recommendations, making further refereeing of a final draft necessary.

   Authors should also be aware that parallel submission of their manuscript to another publisher while under consideration for LNM will in general lead to immediate rejection.

3. Manuscripts should in general be submitted in English. Final manuscripts should contain at least 100 pages of mathematical text and should always include

   – a table of contents;
   – an informative introduction, with adequate motivation and perhaps some historical remarks: it should be accessible to a reader not intimately familiar with the topic treated;
   – a subject index: as a rule this is genuinely helpful for the reader.

   For evaluation purposes, manuscripts may be submitted in print or electronic form (print form is still preferred by most referees), in the latter case preferably as pdf- or zipped psfiles. Lecture Notes volumes are, as a rule, printed digitally from the authors' files. To ensure best results, authors are asked to use the LaTeX2e style files available from Springer's web-server at:

   ftp://ftp.springer.de/pub/tex/latex/svmonot1/ (for monographs) and
   ftp://ftp.springer.de/pub/tex/latex/svmultt1/ (for summer schools/tutorials).

Additional technical instructions, if necessary, are available on request from lnm@springer.com.

4. Careful preparation of the manuscripts will help keep production time short besides ensuring satisfactory appearance of the finished book in print and online. After acceptance of the manuscript authors will be asked to prepare the final LaTeX source files and also the corresponding dvi-, pdf- or zipped ps-file. The LaTeX source files are essential for producing the full-text online version of the book (see http://www.springerlink.com/openurl.asp?genre=journal&issn=0075-8434 for the existing online volumes of LNM). The actual production of a Lecture Notes volume takes approximately 12 weeks.

5. Authors receive a total of 50 free copies of their volume, but no royalties. They are entitled to a discount of 33.3 % on the price of Springer books purchased for their personal use, if ordering directly from Springer.

6. Commitment to publish is made by letter of intent rather than by signing a formal contract. Springer-Verlag secures the copyright for each volume. Authors are free to reuse material contained in their LNM volumes in later publications: a brief written (or e-mail) request for formal permission is sufficient.

Addresses:
Professor J.-M. Morel, CMLA,
École Normale Supérieure de Cachan,
61 Avenue du Président Wilson, 94235 Cachan Cedex, France
E-mail: morel@cmla.ens-cachan.fr

Professor B. Teissier, Institut Mathématique de Jussieu,
UMR 7586 du CNRS, Équipe "Géométrie et Dynamique",
175 rue du Chevaleret
75013 Paris, France
E-mail: teissier@math.jussieu.fr

For the "Mathematical Biosciences Subseries" of LNM:

Professor P. K. Maini, Center for Mathematical Biology,
Mathematical Institute, 24-29 St Giles,
Oxford OX1 3LP, UK
E-mail : maini@maths.ox.ac.uk

Springer, Mathematics Editorial, Tiergartenstr. 17,
69121 Heidelberg, Germany,
Tel.: +49 (6221) 4876-8259

Fax: +49 (6221) 4876-8259
E-mail: lnm@springer.com